T-34 Tank

1940 to present (all models and variants)

COVER IMAGE:
T-34/76 tank. *(Mark Rolfe)*

© Mark Healy 2018

All rights reserved. No part of this publication may be reproduced or stored in a retrieval system or transmitted, in any form or by any means, electronic, mechanical, photocopying, recording or otherwise, without prior permission in writing from the Publisher.

First published in January 2018
Reprinted in May 2022 and April 2024

A catalogue record for this book is available from the British Library.

ISBN 978 1 78521 094 5

Library of Congress control no. 2017947682

Published by Haynes Group Limited,
Sparkford, Yeovil, Somerset BA22 7JJ, UK.
Tel: 01963 440635
Int. tel: +44 1963 440635
Website: www.haynes.com

Haynes North America Inc.,
2801 Townsgate Road, Suite 340
Thousand Oaks, CA 91361

Printed in India.

Commissioning editor: Jonathan Falconer
Copy editor: Michelle Tilling
Proof reader: Penny Housden
Indexer: Peter Nicholson
Page design: James Robertson

Acknowledgements

As with any book, the author has many people to acknowledge and thank for their time, expertise, permissions, photographs and kindness that they made available for use in this work. These are not in any particular order.

In the first instance, to my good friend, Thomas Anderson, for the provision of the German images from his own collection. I also wish to thank Nik Cornish at Stavka.org.uk for the supply of images from his own collection and also the provision of those from the Central Museum of the Armed Forces, Moscow, and the *fonds* of the RGAKFD in Krasnogorsk. I also wish to express my appreciation to Gennady Sloutski and Gennady Petrov for other images from Russian sources. Thanks are also due to the Archives Department of the Tank Museum, Bovington, and also to the museum's photographer, Matt Sampson, for the pictures he took for the walk-around section of the book. My appreciation is also extended to Dick Taylor for the pictures of the Shrivenham T-34/85. For the image of the T-34/76 in the Finnish Military Museum I wish to thank Mr Jari Mäkiaho. I am grateful for permission from Schiffer Publications to use parts of a table in T. Jentz's Vol. 1 of *PanzerTruppen*, and Pen and Sword Publishing for the use of a number of graphic images from M. Baryatinsky's book on the T-34/76 as well as a number of quotations from their publication *T-34 in Action*. I wish to give my thanks, nonetheless, to Yevgeny Scharov for the picture of the T-34/85 and SU-100 in Yemen, even though I was unable to contact him prior to going to press. Thanks also go to J. Vollert for provision of a number of images used in the text. Last but not least, many thanks for the technical assistance from Michael Lane. Other images are from the author's own collection.

T-34 Tank

1940 to present (all models and variants)

Owners' Workshop Manual

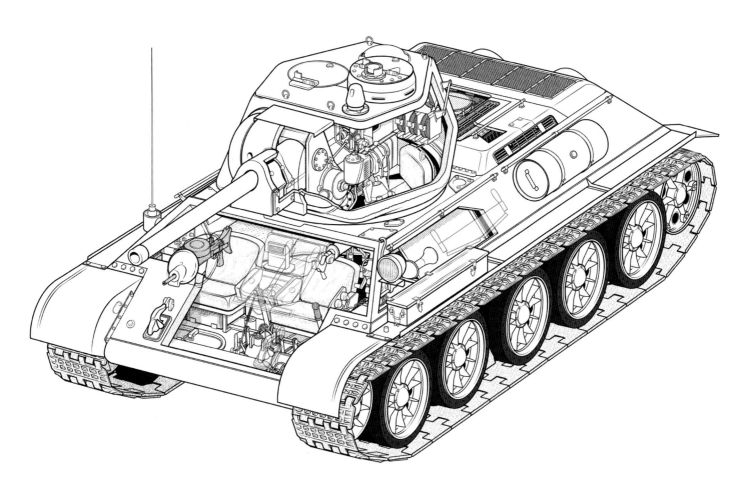

Insights into one of the most influential tank designs of the 20th century and the mainstay of Soviet armoured units in the Second World War

Mark Healy

Contents

| 2 | Acknowledgements |

| 6 | Introduction |

| 12 | The T-34 story |

Spanish and Mongolian lessons 21
The quest for a tank with 'shell-proof' armour 24
Testing times 30
The T-34 that never was – but might have been 33
Order, counter-order, disorder 34

| 36 | T-34 at war 1941–45 |

The cutting edge of Operation Barbarossa 38
The Battle of Mtsensk 48
1942 – the year of 'Deep War' 57
The Battle of Kursk–Orel 71
Operation Citadel, 5–17 July 1943 73
The need for a T-34 with a bigger gun 76
Into combat – the T-34/85 in 1944–45 87

| 98 | Operating the T-34 |

| 104 | T-34s in post-war foreign service |

The T-34/85 in combat: 1950–2017 106

| 114 | Anatomy of the T-34/76 |

Hull 116
Turret 121
Armour thickness and quality 122
Suspension 122
Engine 131
Systems 134
Transmission 137
Towing and lifting eyes 141
Wireless and intercom 141

| 144 | T-34 weaponry and firepower |

T-34/76 with the L-11 76.2mm gun 146
Armament of the T-34/85 146
T-34 with the F-34 76.2mm gun 148
The F-34 in detail 149
Co-axial machine gun 151
Sights 152
Ammunition 153

| 156 | Appendices |

1. T-34 variants including SPGs 156
2. The T-44 158
3. T-34 turrets 160

| 162 | Select bibliography |

| 163 | Index |

OPPOSITE A T-34/85 of the East German Army on parade for the Day of the National People's Army on 1 March 1957. *(Bundesarchiv Bild 183-44519-0001)*

Introduction

This book was started in the late summer of 2016, with the 100th anniversary of the first use of the tank in combat by the British Army at the Battle of Flers-Courcelette on 15 September 1916, a mere week away as the author penned these words. Since that date in 1916, literally hundreds of different designs of tanks have emerged on the world stage and the total of the numbers produced runs into the hundreds of thousands. However, not all tank designs conceived or produced during this period were a success. Many were failures, some were mediocre, while others were effective and were thus produced in large numbers. Very few can claim the mantle of being both revolutionary and to have also made a decisive contribution in determining the outcome of a great conflict. But it is generally conceded that the Russian T-34 medium tank was perhaps the greatest and most important tank ever built.

The use of the word 'perhaps', is intentional. Whereas there are those who would claim otherwise – so the judgement can in no way be definitive – it is nevertheless the contention of this author that the T-34, for all of its faults, and they were many, sits at the top of the list. It is the purpose of this book to try to explain why. As such there will be no attempt to embellish its reputation – it will be presented 'warts and all'. While it was certainly trumpeted by the Soviet authorities as a paragon among armoured fighting vehicles, that was for propaganda purposes, and it is not the task of this work to echo those sentiments if they were not true. On the other hand, it is understandable, given the dire situation of the Soviet Union in the first two years following the German invasion of Russia of June 1941, that this weapon was portrayed as being superior to that of its opponents and was a major factor in helping to stave off defeat. It was certainly the case that the appearance of the T-34 (and the heavy KV tank) in the very first days of the conflict on the new Eastern Front was without question a massive and profound shock to the invader.

That the unquestioned superiority of the T-34 over its German opponents did not do more to impact on the German advance into the Soviet Union did not detract at all from the manner in which the design of all future German armoured fighting vehicles drew on the influence of that of the T-34. Equipped with a new turret and more powerful gun, this remarkable design was the staple of the Red Army tank forces that drove into Berlin in April 1945.

Nor did its career end there. As well as being produced in Polish and Czech factories after the war, it became the primary initial tank exported by the USSR as military aid in the early years of the Cold War. Indeed, such has been the longevity of the T-34/85 that it is still being seen in action in news reports from Yemen and Syria in 2017.

No better summary of the effectiveness of the T-34 can be offered than that of the final paragraph of the British Army School of Tank Technology assessment of a T-34/76 supplied to them for evaluation by the Soviet government in 1943. This document is referred to extensively in this work and it is remarkably objective in its assessment of the weaknesses and strengths of the design. There is no doubting the admiration that the evaluating team expressed for the T-34. In their words:

The design shows a clear headed appreciation of the essentials of an effective tank and the requirements of war, duly adjusted to the particular characteristics of the Russian soldier, the terrain and the manufacturing facilities available. When it is considered how recently Russia has become industrialised and how great a proportion of the industrialised regions have been overrun by the enemy, with the consequent loss or hurried evacuation of plant and workers, the design and production of such useful tanks in such great numbers stands out as an engineering achievement of the first magnitude.

To see, however, how it was that the T-34 came to be, we must return briefly to the beginnings of tank development in the Soviet Union just two decades earlier. It must be borne in mind that this is a massive subject and limited space is available in which to sketch the salient features of the pre-war development of the Soviet tank arm as it pertained to the emergence of the T-34 in 1940.

Birth of the tank in Communist Russia

Within months of the Bolshevik Party's seizure of power in Russia in October 1917, a ferocious and pitiless civil war had broken out between it and those other groups within the country that sought to destroy the new revolutionary government. Aided by arms deliveries from intervening foreign powers that also sought the same end, the tanks supplied by the British and French to the White armies became the foremost symbols in Bolshevik propaganda of the forces seeking to destroy the revolution. From the moment of its birth, the RKKA *Roboche Krestyanskaya Armiya* – the 'Workers' and Peasants' Red Army' (hereafter abbreviated to just Red Army) – found itself embroiled in combat with examples of the latest technology from the battlefields of the First World War. That ultimately the tanks deployed by the White forces did not succeed in helping defeat the Bolsheviks in no way detracted from the impression that these new weapons made on the Red Army, nor how, post-war, they might be employed to their future advantage.

As early as May 1920, G. Sokolnikov despatched a memo to Lenin and Trotsky advocating that even as the Red Army prepared for war with Poland, consideration should be given to the 'utilisation of tanks'. He was referring to nearly 100 British Mark Vs and French FT-17s that had been captured from and later abandoned by the White forces before their defeat and he argued that they should be refitted and prepared for use 'in the formation of tank squadrons'. Indeed, the 1st Tank Detachment was set up using British Mark Vs. Nor were the Soviets dilatory in their attempts to do more than just operate captured machines, for in that same year they produced their first indigenous machine. Based upon the captured French FT-17 light tank, the 'KS' tank (the initials standing for its plant of manufacture, the Krasnoe Sormovo factory in Nizhny Novgorod) was built in 1920 and given the grandiloquent title of 'Freedom Fighter Comrade Lenin'. The leader of the Party was so pleased with it that he personally commissioned 15 more, and in February 1922 they were paraded in Red Square daubed with stirring titles to inspire the people such as 'Red Champion', 'Proletariat', 'The Paris Commune' and 'Victory'.

Although the decade of the 1920s saw energy being expended towards the establishment of the technical means to set up a tank industry – this being paralleled by the development of doctrine for the organisation and employment of armoured fighting vehicles (hereafter AFVs) – this period also saw the growing influence of Germany in these fields in Russia. In the aftermath of the First World War and as a consequence of the resulting peace treaties, Germany and Russia were treated as pariah states. By default it seemed almost inevitable that they would be drawn to one another as a means of circumventing the constraints placed upon them. Almost from the moment the Versailles Treaty was signed, the German Reichswehr and the Weimar government had been casting about in secret for ways to deceive the Allied powers and develop forbidden military technologies. This saw a secret turn to Holland, Finland and Sweden, where through the use of front companies and the willing participation of native arms companies, Germany was able to continue development of U-boats and artillery, and with the Soviet Union, military aviation and tanks.

For the Russians, notwithstanding the recent conflict with Germany that had led to the demise of the former *ancien régime*, Germany and the Germans were still regarded by Russians of all political persuasions as a people of *kulturney* (culture) from whom the new Soviet state and especially the Red Army, could learn and acquire much. And while many in the politically right-wing Reichswehr held their noses at the prospect of co-operation with the despised Bolsheviks, they were nonetheless prepared to engage and collaborate with them to their own advantage. But more than this, they were at one with the Soviets in their mutual detestation of

RIGHT **The first totally indigenous tank design created by the GUVP was the MS-1 (T-18) which emerged in 1927. Formally adopted by the Red Army in 1928 and put into production as an infantry support tank, 960 would be built before production ended in 1931. This figure included the later MS-2 and MS-3 light tanks. This preserved example can be seen in the museum of military equipment at Verkhnyaya Pyshma in Russia.** *(Shutterstock)*

that creation of the Versailles Treaty – the newly reconstituted state of Poland lying between them. Such mutual coincidence of need led to the Treaty of Rapallo in April 1922. This saw the Soviet state grant a concession to the German firm of Krupps to establish a secret tractor (tank) station at Rostov-on-Don. It was here that they developed light and heavy tractors. The Soviet presumption was that, in exchange for this provision, the Germans would share their research and technology to the benefit of their own tank development.

In the meantime, home-grown Soviet tank development stuttered forward – its ability to design new machines was profoundly hampered not by a shortage of financial resources but owing to a lack of expertise and an industrial base broad enough to produce them. Although by the end of 1922, a further 15 of the 'KS' tanks had been produced, the so-called 'Russkiy Renault' was to remain the basic tank in the Red Army through to 1929. In 1923, the Main Department of the War Industry (GUVP) had addressed the problems of developing tanks in the Soviet Union and had constructed a programme based upon the following objectives:

- To carry out all systematic trials possible in the economically underdeveloped Soviet Union.
- To produce equipment for training tank personnel.
- To study tank technology.
- To design and test experimental new models.

While the GUVP was to oversee a number of abortive projects, it nonetheless sponsored an experimental prototype 5-ton light tank designed for the infantry support role. The influence of the earlier FT-18 (based on the KS tank) and the KS tanks themselves could clearly be seen in its design. Following modifications to eliminate problems, it was to emerge as the MS-1 and would be adopted by the Army as the T-18. Production of the T-18 began in Leningrad in 1928 and in its later-developed models, the MS-2 and 3, continued through to 1931, by which time some 960 of differing variants had been produced. It was also in the quest for a tank that could be employed in a

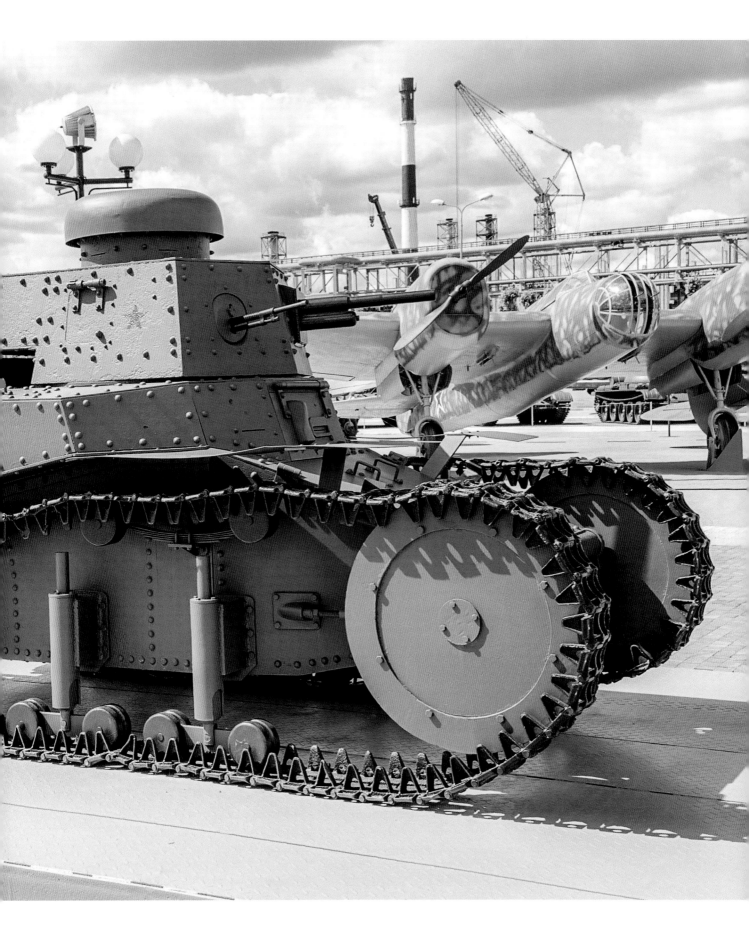

RIGHT The first design to emerge from the then Kharkov Locomotive Works was the T-12. This was the first attempt to produce a 'manoeuvre' tank. Weighing in at just under 20 tons, it was armed with a 45mm gun and three machine guns but it was rejected as being 'mechanically unreliable'. *(Gennady Petrov)*

more independent role within the 'manoeuvre groups' that were now part of the emerging doctrine within the Red Army, that the T-12 tank was to appear.

From 1928 to 1929, the newly established tank design team at the *Paravosostroitelniy Zavod* (the Kharkov Steam Locomotive Works, known by the initials KhPZ) turned their attention to the development of this new machine. In the bureau were individuals such as A.A. Morozov and M. Tarshinov, whose efforts would within eight to ten years help bring to fruition the revolutionary T-34 medium tank. The sole T-12 was trialled in February 1930, but it was deemed to be mechanically unreliable and was therefore abandoned. It did, however, lay the foundation for the improved T-24. This tank was intended to be a replacement for the MS series and while it was slightly more successful than its predecessor – it too weighed 19 tons and was quite well armed, mounting a 45mm main gun and three machine guns – only 25 were built, and as with the T-12, it was ultimately abandoned as being far too mechanically unreliable. Those few nonetheless found their way into military parades and were also used for training purposes. However, all was not wasted as the chassis and suspension were retained and incorporated into the Komintern full-tracked artillery tractor that was built at the KhPZ and used in some numbers by the Red Army through to 1945.

The year 1926 had seen the establishment of a joint German/Russian Tank School at Kazan where experiments with tanks – mainly those captured in the Civil War – were conducted. This site was to become the most important for the interface between the Reichswehr and the Red Army in the area of tank development and although the Soviets would benefit from the interaction, it would not be in those areas in which they desperately needed the practical help they were looking for from the Germans. The Germans secretly shipped in sections of examples of prototype 18-ton tanks from Krupps and Rheinmetall to Kazan in 1928. Various other prototype machines were despatched there, but as one author noted, they were rather choosy as to what they sent:

. . . the Germans did not despatch any of their armoured vehicles which really interested them, such as their early test-bed tractors for the Maybach transmissions. Quite an important factor when one considers that the most advanced and complicated aspects of tank technology are that of steering and gear-changing both of which depend on the transmission system. It was a long time before the Russians were able to develop a satisfactory transmission for tanks.

But access to that sort of technology was exactly what the Russians were hoping to get from their relationship with the Germans. As we shall see, the transmission was the primary ongoing weakness of the T-34, but that machine was still nearly a decade away.

ABOVE **From the abortive T-12 emerged the T-24. Although a more refined machine than its predecessor, its production was curtailed after just 24 had been produced and for the same reason the T-12 had been cancelled – it too was mechanically unreliable. Nonetheless, its suspension was employed in the successful Voroshilovets artillery tractor.** *(Gennady Petrov)*

However, the arrangement with the Germans was to prove more profitable to the Red Army from the standpoint of the development of theory, as from it emerged a 'Correct Line for the War Doctrine of Tanks'. This was tested in practice by employing those same British and French tanks captured in the Civil War in an experimental unit, the results of which were sufficiently promising to lead to the creation of the Red Army's first mechanised brigade in Moscow in 1931.

Nonetheless, the inspiration for the creation of the Red Army's first Mechanised Corps the following year had more to do with acting upon reports of the Experimental Mechanised Force that had been set up in Great Britain in 1927 and renamed the following year as the 'Experimental Armoured Force'. This was the first proper armoured formation ever established in the British Army, notwithstanding the organisation of the Tank Corps in the First World War. Indeed, it was the very first to be set up anywhere in the world and as such it was a milestone in the history of tank development by providing the first occasion in which there was any attempt to create a combined arms force. Comprising of tanks, wireless tanks, tankettes, armoured cars and infantry carried in half-track lorries, it also was fielding its own artillery towed by full-tracked tractors and half-tracks and supporting engineers, again carried in wheeled vehicles. It was unquestionably the progenitor of the all-arms formation, and with Great Britain still viewed by other militaries as the foremost nation in the world when it came to tank development and mobile warfare, this experiment was observed and studied with great interest by those in other countries concerned with how this still 'new' weapon could be best employed in their own armed forces. Although the experimental force was quite limited in size, it nonetheless prefigured the combined arms warfare that was to come. But it would not be the British Army that would build on the fruits of the experience garnered, but rather the Germans and – to a lesser extent – the Soviets. For the latter, it resulted in the expansion of the mechanised brigade into the Red Army's very first mechanised corps under the command of Colonel K.V. Kalinovski, a year later, in 1932.

It was, however, native thinking that saw its doctrinal underpinning in an emergent new concept of armoured warfare called 'Deep Battle'. Its sponsors were a group of radical military thinkers in the Red Army, led by Mikhail Tukhachevsky but also including K. Kalinovskii and V. Triandafillov. In the papers they penned, they envisaged that the future Red Army would be wedded to an offensive strategy in which masses of tanks would be employed to penetrate deep into and behind enemy lines. Only in 1928 with the beginning of the First Five-Year Plan and its commitment to a massive programme of industrialisation primarily to service the expansion of the Red Army, was it possible for them to envisage the state in the near future producing the requisite number of machines to service their ambitious 'vision'.

Chapter One

The T-34 story

When the T-34 appeared in 1940 it was unquestionably the most advanced medium tank in the world. Its sloping armour, wide tracks and heavy gun provided the Red Army with a weapon of huge potential, such that when encountered by the Germans in the summer of 1941 it generated a massive technological shock that influenced all their future tank development.

OPPOSITE This is one of the many T-34/85 tanks that survive in private collections and in military museums. Some, like this T-34/85, are maintained in running order and are demonstrated to the public. In Russia, a number quite fittingly – given the immense contribution made by the T-34 – lead the annual military parade though Red Square that celebrates the victory over Nazi Germany in the Second World War. *(Shutterstock)*

RIGHT General (created Marshal in 1935) Mikhail Tukhachevsky was the most influential theorist in the Red Army and propounded the doctrine of 'Deep Battle' which was based upon the mass use of tanks. His influence was especially formative in the early 1930s until the accusation of plotting to bring down the Party saw his arrest, conviction and execution in 1937. Although his ideas on tank warfare were suppressed, they were employed in practice by the Red Army from 1943 onwards, with the principal instrument of 'Deep Battle' being the T-34. *(Author)*

As 1930 and the new decade dawned, Mikhail Tukhachevsky put pen to paper and waxed lyrical about how the Red Army was on the cusp of realising his concept of 'Deep Battle' on the back of the industrialisation wrought on the basis of the First Five-Year Plan begun in 1928. The case he was advocating hinged on what he saw as the ability of an apparently burgeoning industrial sector to produce the vast number of machines needed to permit the Red Army to fulfil his transformative vision of the future of war. In a paper circulated to Voroshilov, the Commissar for War, and Shaposhnikov, the Head of the Red Army Staff, plus others, he set forth what he saw as the material basis underpinning his case in

> . . . the successes of our socialist construction, the accelerating tempo of our country's industrialisation, and the socialist reconstruction of agriculture set before us in all its magnitude the task of reconstructing the armour forces taking into account the newest factors of technology, the possibility of mass military-technical production.

Such became his rationale for his advocacy that the means now existed to expand the number of tanks in service with the Red Army by 1932–33 to the immense figure of 50,000 machines! Furthermore, he also argued for a parallel expansion of airpower, all of which would combine and serve to permit the Red Army to conduct a future war.

As Head of the Red Army Staff, it fell to Shaposhnikov to subject the document to rigorous scrutiny, in consequence of which he proceeded to pen a lengthy riposte to Tukhachevsky's thesis. While acknowledging the 'radical' extent of the ideas expressed, he nonetheless grounded his critique in the supreme reality facing the Soviet government at the time by saying: '. . . neither the nation nor the economy would be able to provide the material and manpower resources that Tukhachevsky believed possible.' Voroshilov passed both documents on to Stalin who was somewhat more scathing in his comments, labelling the content of Tukhachevsky's paper tantamount to 'science fiction'. His most telling observation was that he wondered how was it that a Marxist commander could separate military matters from the country's underlying economic realities when advocating his case. While Stalin continued to nurse a personal antipathy to Tukhachevsky that extended back to 1920 and which is certainly discernible in the scorn of his put-down of the general, it is also the case that his observation also serves as an oblique reference to the problems being experienced in the programme of industrial development under the First Five-Year Plan at the time. Not least of these were those concerned with tank production, the reality of which totally ridiculed Tukhachevky's pretensions. Nor could Tukhachevsky, as the Commander of the Leningrad Military District, have been unaware of these as the city was home to a large number of armaments works, of which three were involved in tank production at this time. Likewise he could not have been totally oblivious to the wider problems facing tank production in the USSR.

How dire the situation was is revealed by industry's failure to achieve the less than ambitious tank-building plans set out for the fiscal year 1928–29 – the first year of the First Five-Year

Plan. A mere 65% of the targeted production had been achieved by March 1930, with the programme for 1929–30 being in a yet worse state, with even the paltry production target of 10 tankettes and 30 T-12 medium tanks not being realised. Even as Tukhachevsky was predicting an army of 50,000 tanks that would overwhelm any aggressor, factories within the Soviet Union had not managed to manufacture one tank for that year's quota! Nor had matters improved by September 1930, when I. Khalepski, as head of the *Upravlenie –Motorizatsii i Mekhanizatskii* (Directorate of Motorisation and Mechanisation – UMM), had to inform Voroshilov 'that one of the most important tank plants was still unable to produce the T-24 tanks according to plan'. This was a reference to the KhPZ plant that was having difficulty mastering the technology. These problems were also mirrored across the wider industrial sector. The Plan itself was in trouble. Although state propaganda lauded the 'great strides' being made in all areas of the economy, in reality it was 'overheating' – a product of excessive and unrealistic goal-setting that took no account of the real deficiencies under which it laboured at this stage in its development. Stalin's response was to lash out, insisting that this was all the product of the work of 'wreckers', who were acting as agents of the capitalists and engaged in mass sabotage. The introduction of punitive measures against managers and workers targeted as such was not, however, reflected in the attitude taken towards those in the tank industry. In this case Voroshilov, no doubt with Stalin's acquiescence, more realistically and objectively attributed the problems in tank production to 'our feebleness' – a recognition that there existed a profound weakness in tank design and in the lack of a basic automobile and tractor industry geared to mass production. This was a matter that required prompt redress if the Red Army was to give any substance at all to its ambition of building a strong and effective armoured force. The issue was, however, how best to deal with the matter.

Consideration was in the first instance given to streamlining the number of tank types being developed to produce just one. But on reflection, this was rejected in favour of acquiring foreign technology to circumvent the problem. It was Khalepski who was delegated to head a special tank commission to go abroad and seek out the expertise that could help the Soviet tank industry surmount its 'feebleness' – in technology generally and in tank design in particular – thereby acquiring the means to address the failures in tank production. In his role as Head of the UMM, he had already advocated the twin-track approach of both developing an indigenous tank industry alongside the import and exploitation of foreign technology to aid in the development of the former. It was this suggestion, given official endorsement by Vorishilov and sanctioned by Stalin that had prompted the Red Army, even before the end of 1929, to enter into discussions with companies in Europe, Great Britain and the USA. This was also a reflection of the growing dissatisfaction of the Soviets with the Germans, which was to build up in the early years of the 1930s. In essence, this decision to cast the net wider was prompted by the desperate need to find the technology and expertise necessary to develop the tank industry.

The reality was that the Red Army had benefited little from the technology the Germans had brought to the Kazan training school. Although the site was controlled by the Germans, as it was paid for by them and was therefore understood to be servicing German interests, it nonetheless irked the Russians that they had seen so little benefit from the relationship. What had been hoped for, above all, was that the Germans would have helped address what Voroshilov called the 'feebleness' in Soviet tank design and technology. As Commissar for Defence he penned a letter to the German General Wilhelm Adam in which he explicitly set down Soviet suspicions about their intentions at Kazan:

> *We already have an industrial base, but we still have very few people – i.e. designers. You have people. We thus thought that your side would furnish models, blueprints, drafts, ideas, designs in a word, that we would have laboratories both for you and us. None of this has happened.*

The same message was communicated by Tukhachevsky in his new role of Inspector of Armaments, to the then German ambassador to Russia, Herbert von Dirksen, when he accused

RIGHT In 1930, the Soviet Union bought examples of the Vickers 6-ton light tank and the rights to produce them. This AFV became the basis for the very large number of the T-26 light tanks. *(Gennady Petrov)*

the Germans of 'holding back supplies and new technology from the Soviet Army'. In this matter both men were correct. The underlying German antipathy about the relationship with the Soviet Union, despite the obvious benefit they were deriving from it, was summed up by a comment by General Kurt von Hammerstein-Equord, the then Chief of the Army Command of the Reichswehr, who likened it 'to a pact with Beelzebub'. The understanding had but a little time to run in any case, as Hitler's accession to power in January 1933 and his very vocal antipathy to the Communist state prompted the Soviets to close down all the sites used by the Germans in the USSR. By September, Kazan had been terminated and the Germans were gone. But by this time the Soviet tank industry was in a different situation to how it had been in 1930. There was a growing confidence that the problem of 'feebleness' had been addressed with solutions sourced not from Germany, but the United Kingdom and the USA.

RIGHT J. Walter Christie was a maverick US tank designer who founded his own corporation to build his creations. He sits here atop the turret of his M1931 Medium Tank T3 as it is being assessed by the US Army, who ultimately bought four for the US Cavalry in 1932. However, the US Army rejected the Christie suspension, unlike the Russian and British armies. *(NARA)*

ABOVE **Although a later variant than that purchased by the Soviet tank commission this T3E2 nonetheless demonstrates the Christie suspension that so impressed Khalepski. Both tanks are seen here in their tracked mode, whereas they could also be employed as a wheeled AFV. The antecedents of the Russian-built BT series is very clear to see.** (NARA)

In late 1929, Khalepski, aided by the tank designer Semyon Ginsberg, had embarked on a tour of major tank producers in Europe, visiting Škoda in Czechoslovakia, Somua, Citroën and Schneider in France and, in early January 1930, Rheinmetall, Krupp, Daimler-Benz, Krauss-Mafei and Linke Hoffman in Germany. Given the links between the Germans and the Soviets it is not surprising that Khalepski's stay was quite long, involving as it did detailed discussions in which he stressed that the intention was to seek assistance in building new tanks and not in developing existing ones. From Germany, the Soviet party travelled to Great Britain, where they visited the Vickers Company. At this time this British AFV manufacturer was regarded as one of the leaders in tank design and development, with both Khalepski and Ginsberg being impressed by what they were shown of the products under development and in production. In consequence, they entered into a commercial agreement with Vickers to purchase 15 of the privately developed 'Six-Tonner' light tanks, which type was to emerge as one of the most significant armoured fighting vehicles of the 1930s, being bought by many foreign nations, though not by the British Army. The purchase of 15 of these tanks went through without any difficulty because it was a commercial transaction, whereas the purchase of the other types that the Russians contracted for took a little longer, as they had been bought by the British Army and the licence for their export was delayed for that reason. In addition to the 'Six-Tonners', the shopping list included 26 Carden-Lloyd tankettes, 8 amphibious light tanks and 15 Medium Mark IIAs. The 6-ton tanks arrived in Russia between September 1930 and January 1931, where they were designated the V-26 – the 'V' standing for Vickers. From Great Britain, Khalepski and his party next ventured across the Atlantic to the New World, there to avail the USSR of the latest products of US tank technology. The primary focus of Russian attention, however, fell on the products of the fertile mind of the maverick tank designer J. Walter Christie who had ploughed a lonely furrow in his development of track-laying vehicles. Although his first products dated back to the beginning of the 1920s, it was on his M1928 design that Khalepski's attention fell. What attracted him was its novel suspension employing identically sized road wheels attached to large helical springs which were mounted within covers, inside the hull. This permitted the M1928 to achieve fast speeds on roads and across country (indeed, as we shall see, the suspension of the T-34 of 1940 was, in its essentials, identical to that of the M1928). s Although the design, as seen by Khalepski and his team, mounted only a machine gun, it was clear that the design had the potential they were looking for. In consequence, and acting on Khalepski's advice, Defence Minister Voroshilov sanctioned the purchase of two Christie tanks and the licence production rights for $60,000. This was followed by a further contract for $100,000, which purchased the patents and production rights to produce and develop the now renamed M1940 in the Soviet Union. It also

ABOVE The T-26A model was the first variant built in the USSR and entered Red Army service in the early 1930s. This example of the T-26 is on display near Leningrad. *(Shutterstock)*

had the facility for the tracks to be taken off and the machine to be driven on its rubber-shod road wheels. It was thus a wheel-cum-track design – a concept that was to retain much 'traction' within the Red Army in the 1930s.

With the purchase of the Vickers 6-ton tanks and the Christie M1928, the USSR had the two machines that in their developed forms would provide the light tanks of the Red Army through to the onset of the Russo-German War. While initially the production of both types was concentrated at factories in Leningrad, the decision to expand tank production beyond just these saw the production of the BT-2 – the first

RIGHT The BT-1 was almost a straight copy of Christie's M1931 wheel-cum-track machine. The BT-2 seen here mounted a 37mm gun and ball-mounted machine gun. BT stood for 'fast tank'. *(Copyright unknown)*

LEFT The BT-5 was developed in 1932 being armed with a 45mm main gun and a three-man crew. This example was captured by the Finns and reworked by them to serve in their own small tank force. This example is in the Break of the Siege of Leningrad Museum at Kirovsk. *(Shutterstock)*

production model – shifted to Kharkov. It was from this date that all further development of the BT series was undertaken in the Kharkov works. This included the BT-2, BT-5, then the BT-7 and lastly the diesel-powered BT-7M. It was built in the thousands to satisfy the need for a fast tank in the Red Army.

BELOW The ultimate version of the BT series was the BT-7 and was developed at the Kharkov Zavod. Producion began at the end of 1934 and 4,613 were built. The last version – the BT-7M – was equipped with the same 500hp diesel engine that would eventually power the T-34. When war began in June 1941 the Red Army had some 7,549 BTs of all models on strength. This BT-7 can be seen at the museum of military equipment at Verkhnyaya Pyshma. *(Shutterstock)*

Neither did the influence of foreign tank design – and especially that of the British – end with just the T-26 and BT series light tanks. This was also seen in Soviet medium and heavy tank design, which were profoundly influenced by the novel and idiosyncratic Vickers multi-turreted 'Independent' tank. Making its appearance in November 1926, it was employed in extensive and ongoing trials. By 1930, a lot of experimental results had been accrued, but all through the period of these tests, those conducting them were never sure whether the Independent was a one-off or a prototype prior to series production. Employing the same concept but mounting three, rather than five turrets, was another of Vickers' many progeny – the Medium Mark III and its developmental offspring. Like its heavy forebear, it too did not result in production orders for the British Army. Both of these types would have their greatest impact overseas.

What made both of these designs of interest to the Soviets as well as the Germans and French, was their novel multi-turreted arrangement. The main turrets of both machines carried a 47mm Quick Firing (QF) gun, while the four subsidiary turrets of the Independent were stationed at the four quarters relative to the main turret, with the two of the Medium Mark III carried on the front of the machine and below the main turret. In both cases, each subsidiary turret mounted a Vickers machine gun. That the notion of a multi-turreted AFV was conceptually flawed and in terms of tank design an evolutionary dead-end, did not prevent other nations from adopting it themselves. In the Soviet Union, the idea of a multi-turreted design helped foster the T-28 medium tank and the heavy T-35. Both machines would become symbols of the armoured might of the Red Army and could be regularly observed trundling their way across Red Square in the military parades held in the 1930s to mark the important days of the Communist calendar. Their sheer size and numerous turrets embodied the might of the Red Army, thereby serving as impressive propaganda 'vehicles' for the Soviet state. The T-28 would also figure frequently in propaganda films extolling the prowess and readiness of the Red Army to defend the *Rodina*.

Unlike its heavier stablemate, the T-28, although multi-turreted, drew on the concept of the Medium Mark III but was the first machine of wholly Russian design. Originating from the Leningrad Kirov Plant, the prototype emerged in 1932. As originally constructed, it mounted a 45mm gun, later to be replaced in the production variants by a short-barrelled 76.2mm weapon in addition to between three and four machine guns – one co-axial and the

RIGHT The T-28 three-turreted tank equipped many medium tank units in the mid- to late 1930s, although they were obsolete by the time of the German invasion. It was armed with a 76mm gun in the main turret with machine guns in the two sub turrets. This example is displayed in Moscow at the Central Museum of Armed Forces. *(Shutterstock)*

ABOVE Inspired by the design of the British 'Independent' multi-turreted tank, the T-34 was the perfect machine for Soviet propaganda. The pictures of these machines trundling through Red Square cemented its image as the Soviet 'land battleship'. *(Shutterstock)*

other two in the secondary turrets. The fourth could be added to the gunner's hatch as an anti-aircraft weapon. Weighing a little short of 20 tons, it was designed for employment in the new tank brigades. A second model appeared in 1933, being designated the T-28A. This had an improved suspension and was to remain in production until it was replaced in turn by the last model, designated the T-28B (the T-28C was the T-28B with extra armour added to the hull front and turrets following the Winter War in 1939). Some 503 of all models of the T-28 would be produced through to 1940 with just 13 being the final number to be manufactured in the same year that the first T-34s left their production line in Kharkov. Although the Red Army had determined that the T-28 was obsolete long before the outbreak of the conflict with Germany, it saw limited and unspectacular service in the Winter War with Finland in 1939/40, and there were still 411 in service as of the beginning of 1941. A few survived to see action of the Leningrad Front as late as 1943.

The first Russian attempt at developing a heavy tank based on or certainly inspired by the design of the British 'Independent' was the T-32. It too had five turrets, arranged in the same manner as the Independent, but was much more heavily armed. It saw limited production before it was replaced by the T-35, the prototype of which emerged in August 1932, although the T-35A, which was built at Kharkov, was sufficiently different to the former to have been regarded as a completely new machine notwithstanding the visual similarities between the two types. The design was problematic, stemming in part from the machine's complexity and though it made quite a spectacle in the Red Square parades of the mid-1930s – being described as a 'land battleship' – production ended in 1939 with just 63 having been built; 61 of these were still in the inventory of the Red Army in June 1941, with the largest number serving in the 8th Mechanised Corps which had 48 on strength. These were all lost, the bulk owing to breakdowns and mechanical failure. Two survived to take part in the Moscow counter-offensive in December and one at least was captured in action on film.

Spanish and Mongolian lessons

The decision to commit armour as part of the forces sent from Russia to support the Republicans in the Spanish Civil War in 1936 marked the first occasion that the Red Army could assess both the effectiveness of its tanks and also see how well the doctrine that governed their operation coped with the actual conditions of combat. This was not the only formative experience they drew upon, for within a short period after the return from Spain, the tanks of the Red Army once again found themselves in action but on this occasion fighting the Japanese Army many thousands of miles from the Iberian Peninsula, at Lake Khasan and Khalkin Gol in Mongolia. The experience and the effectiveness of the T-26 and BT light tanks that served in both theatres and the post conflict analysis applied

ABOVE A T-26 seen here on display in Volgograd. Along with the BT series the different variants of the T-26 made up the bulk of the strength of the Red Army tank force throughout the 1930s. This type saw service in the Spanish Civil War. *(Shutterstock)*

to the reports of their operations served not only to impact on the continuing development of doctrine within the Red Army, but more importantly for our concerns, fed into the specifications for new tank designs from which would in due course emerge the T-34.

In total, 106 T-26 and 60 armoured cars were sent in 1936 followed by 50 BT-5 tanks and a further 175 T-26s in 1937. There was no question that the two Russian types showed themselves to be superior to the poorly armed Panzer 1 machine-gun-armed tanks employed by the Germans. The 45mm gun of the two Russian types permitted them to 'spot the German vehicles, and shoot from up to a kilometre away and yet penetrate the thin armour of the smaller tanks'. Nonetheless, the poor handling of the Russian tanks by Nationalist commanders reduced their combat potential. This was reflected in the tank losses sustained by them which were deemed to be quite high – in consequence of which the Soviets were to derive some questionable judgements about their future employment. In this they were at variance with those drawn by the Germans, wherein the general consensus was not to attribute such significance given the demonstrated inadequacy of the Panzer 1 and the limited scale and nature of the conflict.

In their post-war analysis, a number of influential Red Army officers took the opposite view arguing that the Spanish Civil War did indeed offer a genuine picture of how a future war would be prosecuted and that it provided an accurate pointer of the role of the tank within it. The primary consequence arising was the manner in which it served to raise serious questions about the continuing validity of the concept of 'Deep Battle'. Indeed, what was now being uttered by some was a notion that ran totally counter to this and would take the Red Army in a direction entirely contrary to that of the Wehrmacht. In place of huge mechanised corps speeding to suppress the entire breadth of the enemy's positions, the lessons of Spain suggested a slow, methodical advance by the machines as they worked to perform their older mission of helping the infantry to move forward. The armoured assault had to be heavily supported by infantry and artillery, since tanks could not move without protecting their foot soldiers and certainly should not attempt to fight independently in the enemy's rear or on his far flanks. The most influential voice advocating this view was provided by General D.G. Pavlov who had served in Spain and reported to Stalin and Voroshilov that 'the tank can have no independent role on the battlefield', leading to the conclusion that the tank battalions be distributed in an infantry support role. Under normal circumstances and in any other state such a fundamental move away from what had been more or less an accepted view of the role of the tank would have generated vigorous professional debate within the ranks, but his opinions had acquired a degree of traction because of the unprecedented assault by Stalin on the leadership of the Army. Whatever the byzantine machinations that lay behind Stalin's decision to purge the Officer Corps of the Red Army, its consequences would prove nearly fatal for the Soviet Union when war broke out with Nazi Germany in June 1941. The first step was taken against Tukhachevsky, who was arrested and accused of fomenting an 'anti-Soviet' plot against the government while acting as an agent of Fascist Germany. On 11 June 1937, a trial was held in camera and was conducted in great haste with no semblance for the notion of justice, for judgment had already been passed. Alongside Tukhachevsky were seven other very senior officers who had been arrested a few days after him. They faced a bench made up of those whom they regarded as friends and colleagues. Following their denials of the charges ranged against them all were subject to 'vigorous treatment', a euphemism for severe torture so as to generate the self-confession required by Stalin with that signed by Tukhachevsky spattered with his own blood. Despite appeals for clemency they were sentenced to death with indecent haste and summarily executed by firing squad in the courtyard of the NKVD building in Dzerzhinsky Street in Moscow on 12 June 1937.

This was merely the overture as the stain of manufactured suspicion spread rapidly to encompass virtually every level of the officer corps. Many officers who had expressed support for the notion of 'Deep Battle' were thus tainted by association and paid the price by being 'repressed' (this included Khalepski). Only those individuals who were perhaps less committed to the notion and who publicly recanted this now ideologically heretical idea survived. The concept of mobile operations – as articulated in the West by the likes of Fuller and Liddell Hart, two British thinkers whose writings on the role of the tank in modern warfare were highly influential in Germany and Russia at this time, and trumpeted by Tukhachevsky and his followers for so long – was denounced as reactionary and traduced by its being labelled a product of a 'decadent capitalism which dared not place its trust in the masses and so hid behind mechanical contrivances'. In essence, the ideological verbiage rolled out to justify the purge of the military was reducible to the formula that mobile warfare was bourgeois and unworthy of a Marxist society. It was noted by one author that:

> *It was no coincidence that the tank troops were purged more severely than other branches of the service, but rather a direct outgrowth of suspicions already levelled at the authors of 'Deep Battle'. The targeting of these men, added to the failures to implement the doctrine, would force a thorough reconsideration of the tank in battle.*

This by default reinforced the case being argued by Pavlov, who with the demise of Tukhachevsky, now came to be viewed by the leadership as the foremost tank specialist in the Red Army and the case he was advocating was in direct opposition to that of 'Deep Battle'. A special commission was set up in 1938 and overseen by another of Stalin's Tsaritsyn cronies, G.I. Kulik – one of the 'bears of little brain' who formed the Vozh's inner circle. It was manned by other leading lights of the Red Army hierarchy, as well as military district commanders, all of whom were extremely concerned to keep their heads and thus mouthed the necessary platitudes, and supported the final conclusions that the armoured corps required 'reforming', and therefore backed the case being advocated by Pavlov. In the face of the dissent of Shaposhnikov and Zhukov – the latter being the victor of Khalkin Gol – Stalin and Voroshilov rubber-stamped Pavlov's views and on 21 October 1939 the order was passed for the dismantling of the mechanised corps and their subsequent reorganisation into 15 motorised divisions. Eight of these were to be created in 1940 and another seven in the first half of 1941. The future task of the tanks was now deemed to be that of infantry support.

It was a catastrophic decision. In a short time, many of those who had long advocated the idea of the tank force in the Red Army as an independent arm were rotting in their graves. The loss of their intellectual expertise and the dismantling of the armoured corps – plus the return of the tanks to a role little different to that employed in the First World War – was in the longer term to directly contribute to the calamity that would be visited on the Red Army barely two years later. Given how huge was the institution of the Red Army and over how vast an area it was deployed, the impact of these changes would necessarily take some time to be initiated and then 'bed in'. No sooner had this decision been made and the changes enacted, however, events abroad would show that it was a profound error.

The quest for a tank with 'shell-proof' armour

It was not just in matters of armour doctrine that the experiences of the tank force sent to Spain served to prompt debate. Although both the T-26 and the BT tanks that had served there had shown themselves superior to their German and Italian counterparts, post-war analysis had led the Red Army High Command to argue that their own tanks were less than adequate for the conditions to be found on the modern battlefield. This perception received further confirmation when these same two types were pitted against the Japanese Army at Khasan and Khalkin Gol in Mongolia in 1938/39. While their main armament was shown to be more than adequate for dealing with the tanks of the Japanese Army, the thinness of the Russian light tanks' armour led to high losses from artillery and other weaponry. In Spain and then in the later fighting in Mongolia, this revealed itself in a tendency for these petrol-powered machines to catch fire rather too easily.

In a post-conflict report, R. Ia. Malinovskii (later a Marshal of the Red Army and one of the few who served in Spain not to be repressed), specified that it was the lack of armour protection that was the primary reason for the high losses among the T-26 and BT tanks. His recommendation was to replace those same two machines with new tanks with much thicker armour. The perception arose that a change in specifications for tanks for the Red Army was needed:

BELOW A large number of BT tanks along with T-26 tanks saw combat with the Japanese Army in 1939 at Khalkin Gol in Mongolia. Although a Soviet victory, the petrol-engined Russian tanks proved vulnerable to Japanese fire. From the experiences of the Spanish Civil War and Khalkin Gol came the need for a 'shell-proof tank'. *(Gennady Petrov)*

The experience gained during the Spanish Civil War, and the trend in anti-tank artillery development during 1936–7, revealed the necessity for considerable alteration on the fundamental combat characteristics of tanks; increases in both armour protection and firepower. . . . It became necessary to introduce 'shell-proof armour' and a substantial increase in firepower.

In spite of Pavlov's other questionable judgements, as head of the ABTU, he did recognise the need for the Red Army to acquire a new tank, the design of which would encompass the requirement to have 'shell-proof' armour. Whatever the thickness of armour that was to be employed, it would need to be able to defeat 37mm anti-tank gun fire at any range and a 76.2mm anti-tank gun at ranges over 1,000m. As it was, the Automobile and Armour Directorate (AAD) had already been charged to 'draft and build, experimental vehicles and put into mass production a fast tank with synchronized wheel-cum-track chassis'. This order had been passed to director Bondarenko of Zavod No. 183 in Kharkov on 28 September 1937. As we shall see, the issue of whether the wheel-cum-track was desirable or even necessary was to become a matter of fractious dispute. In the first instance, this new design was to draw upon the work done to improve the earlier BT series. A separate design bureau was set up within the Kharkov factory to work alongside that already engaged in updating the BT tank, with this new design being allocated the designation BT-20. Thus began an evolutionary design process which would lead, within a few years, to the emergence of the T-34. Design leadership for the BT-20 was given to M.I. Koshkin, for which he established within Plant 183 a specialist team of 21 technicians within a sub-department of the design bureau numbered KB-190 to address the matter.

The name of Mikhail Koshkin is bound up with what has been described in some quarters as the 'myth of the creation of the T-34'. There is no question that Koshkin was an extremely able designer with a strong sense of self-belief. That much must be inferred by the manner in which he was 'head hunted' – to use the modern expression – by Sergo Ordzhonikidze, the then People's Commissar for Heavy Industry, who sent him to Kharkov in the place of the previous Head Designer of Zavod No. 183, who had been 'repressed' in the Purges. Koshkin's career from the time of his obtaining a degree in engineering at the Moscow Communist University in 1924 is an interesting one. The first application of his engineering expertise was as an apprentice in a sweet factory in the town of Vyatka. Convinced that he could achieve better things, he returned to school. His more than satisfactory record as a young Bolshevik secured a place at the prestigious Leningrad Polytechnic Institute where he specialised in car and tractor design, for which he showed a remarkable aptitude in a department that was itself known for fostering high-flyers. Graduating in 1934, he became a specialist in tank design and within two years his contributions in this field had resulted in the award of the Order of the Red Star. His ability was without question and it was this above all else that probably prompted Sergo Ordzhonikidze to promote him to Chief Designer at Zavod No. 183, at the fairly young age of 38.

Koshkin is often credited with the first ever use of sloped armour on the A-20 tank, but

LEFT The name of Mikhail Koshkin – the designer of the T-34 – is revered in Russia to this day. His life story was made into several films in the former USSR and his image, as seen here, was to be found on a number of Soviet-era stamps. He was quite young when appointed to oversee the KhPZ and was behind many of the designs that emerged in that Zavod before his untimely death in 1940. *(Author)*

ABOVE **Although not the first tank to have sloped armour, the BT-IS was the ultimate expression of the BT series of tanks. It did not go into production.**
(Copyright unknown)

in truth, no matter how able he was, it was another who introduced this feature into Soviet tank design. N. Tsyganov had introduced the idea of sloped armour into Russian tank design in his BT-7-IS tank – itself a development of the earlier BT-7 – and the ultimate development of the BT series. It is probable, however, that Tsyganov was aware of developments abroad, as the very first tank to be designed with sloping armour was the French FMC-36 infantry support tank, which had entered service with the French Army in 1936. The sloped armour he created for this new variant of the BT-7 prompted the nickname for his creation of 'Cherepakha' (or 'Tortoise'). It was more formally titled 'Ispitatelniy' (meaning 'Investigator'). The Soviet tank historian Mostovenko wrote:

> *In this vehicle the method was tested of constructing an armoured hull in which all the armour components on the front, sides and rear were greatly inclined to increase the immunity from shell fire. The turret was constructed along the same principle.*

The official designation was as the BT-IS (with the 'IS' part standing for Iosif Stalin). While it was always sound policy to name a tank or some other vehicle after a leader, in Tsyganov's case it was to no avail, as he too was 'repressed' as were others in the Kharkov design bureau in 1938. Although his design for the BT-IS died with him, the efficacy of the sloped armour used on Tsyganov's design was recognised and taken up by Koshkin and employed on all of the early development machines from the BT-20 onward, preceding the emergence of the T-34 in 1940.

The requirement for the BT-20 design was detailed and specific. But at this stage Pavlov only saw the BT-20 as a replacement for the BT series with the new design drawing heavily on the BT-IS. As such he saw it as another, though superior, light tank. That it was to mutate into a larger, heavier and more powerfully armed medium tank was not of his doing, nor was he supportive of it. In the light of how important the BT-20 was to be (as the evolutionary ancestor of the T-34), the specification handed to Bondarenko as Plant 183's director and thence on to Koshkin, is worthy of examination. It was as follows:

- **Type** – wheel-cum-track, 6-wheel drive, Christie type.
- **Combat weight** – 13–14 tons.
- **Armament** – 1 × 45mm 3 DT gun and a self-defence flame-thrower. Every 5th tank must have a flame thrower or, 1 × 76mm 3 DT gun and flame-thrower. Every 5th tank must have an anti-aircraft mount.
- **Ammunition load carried** – 2,500–3,000 rounds MG 130–150 × 45mm or 50 × 76mm rounds.
- **Armour:** front – 25mm, conical turret – 20mm sides, rear – 16mm, roof, bottom – 10mm. The armour should be all sloped, with a min angle of 18 degrees for the slanted hull and turret plates.
- **Speed** – the same for track and wheel running gear: max 70 km/h, min 7km/h.
- **Crew** – 3.
- **Cruising range** – 300–400km.
- **Engine** – BD-2 developing 400–600hp.
- **Transmission** – BT-IS wheel-cum-track type (driving power when on wheels taken from steering clutches).
- **Suspension** – independent; torsion bars preferable over carriage springs.
- Orion stabiliser and turret traverse stabiliser by engineer Povalov; searchlight for night firing, reaching out to 1,000m.

It took until March 1938 for the BT-20's design drawings to be approved by the AAD. Now designated as the A-20, the next hurdle to be surmounted was the meeting of the State

Defence Committee (GKO) on 4 May. Very much a congregation of the Soviet great and good as related to defence matters, it was chaired by the Soviet premier Molotov and included Stalin (whose authority, even though he was not chairman, was pre-eminent), Voroshilov and a large number of leaders from the military, the Party and defence industry. Stalin expressed the general view to the designers present that he wanted 'superiority over Western tank designs in firepower, armour protection and mobility and insisted upon an increased capacity for long-range employment' – sentiments that were music to both Koshkin and Morozov's ears.

Koshkin's team presented the work carried out thus far on the A-20 with the aid of a wooden model and it was while doing so that he ventured to express a view that ran counter to that held by the majority in the meeting: that his team had doubts about the need for the A-20 to be a wheel-cum-track machine. He argued that not only did the mechanism to permit wheeled mode function add unnecessary weight, but that the evidence of recent combat experience had shown that this dual automotive facility was not necessary and that a wholly tracked vehicle would be more effective. This was provocative stuff and a vigorous debate ensued during the course of the meeting as to the respective merits of both. Those supporting the case for the former insisted that a tank travelling over long distances would be better placed to employ wheels before resorting to fitting the tracks prior to going into combat. The minority countered by arguing that the issue was not about reliability; rather it was whether the complex mechanism required for the suspension to permit a tank to be in one mode or the other was necessary at all. It was their contention that the A-20 would, for those reasons, be more complex to produce and that this would also serve to raise its cost and slow its speed of production.

While Stalin generally held most of those around him in little regard and had no compunction in despatching even close associates and relatives to the Gulag or even worse, he nonetheless was prepared to give his attention to those who stood their ground and argued their case in his presence. It must be presumed that Koshkin looked him in the eye when making his assertions – something that Stalin viewed as some sort of 'litmus test' as to an individual's veracity. The Vozhd lent his ear as Koshkin seized the moment and, going beyond his immediate brief, put forward the case for a tank that was not constrained by the specification for the A-20. He stated that to contend properly with future threats the armour of a new tank needed to be at least 30mm thick. Furthermore, the 45mm gun with which the A-20 was to be armed was inadequate and that a 76.2mm weapon would be needed to defeat the more heavily armoured types that were bound to emerge. Stalin was impressed by what Koshkin said and, following his lead, the members of the State Committee sanctioned the building of prototypes of the A-20 and also of another machine to be designated the A-30, which was to have heavier armour and a more powerful main gun.

Returning to Kharkov, Koshkin's team discovered that subsequent to the 4 May meeting the AAD had issued a further set of instructions modifying the A-20 design. Although it was to remain a wheel-cum-track machine, its armour had been increased so that it could now withstand hits by armour-piercing 12.7mm heavy machine-gun fire, prompting the armour to be cast at a more acute sloped angle than originally called for. The A-20 was given a wide glacis plate set at a 60-degree angle. The hull overhung the tracks with the angled armour being 25mm thick. This was strongly reminiscent of the arrangement previously adopted for the abortive T-111 (T-46-5) design of 1937–38. Although this had not proceeded

BELOW The T-111 (T-46-5) was only a prototype but was important as being the first tank in Russia built to have 'shell-proof' armour. It was rejected in favour of the KhPZ-designed A-20 and A-30. *(Copyright unknown)*

RIGHT A.A. Morozov worked alongside Koshkin in the design process that led to the T-34. An engineer of great ability, he was responsible for the design of the diesel engine that was to power it and many other Soviet AFVs. He took over the leadership of the design bureau on Koshkin's death and oversaw its continued work when it was evacuated to Nizhne Tagil in late 1941.
(Copyright unknown)

BELOW An A-20 tank in wheeled mode under test at the Kubinka Proving Grounds outside Moscow in 1938.
(Gennady Petrov)

beyond the prototype stage, the experience accrued in the design and building of this machine was not lost, as it was recognised that vital experience gained in the construction of a tank with shell-proof armour could be employed in new, future designs.

And this experience, in terms of the expertise of a number of the engineers who had worked on the T-111 project, was embodied in several members of Koshkin's 'team' working on the A-20. Foremost among these was A. Morozov, who would subsequently become Koshkin's successor as Head of the Design Bureau. A man of great ability, he had been the designer of the revolutionary V-2 diesel engine that would power the A-20 and all subsequent designs including the T-34. It was already being utilised on the very last variant of the BT series being built at the Kharkov Zavod – the BT-7M. He and M. Tarshinov, who had worked alongside N. Tsyganov on the BT-IS, were allocated the responsibility for overseeing the design of the hull. Two other members of the 'team', N. Kuchereneko and P. Vaishev, who had been party to the design of the suspension of the T-111 (a novel attempt to adapt the Christie-type suspension to a medium tank), were tasked with leading the suspension team on the A-20.

Nailing their colours to the mast, Koshkin and Morozov submitted a paper to the Soviet High Command in which they stated:

In view of the tactical reluctance to employ the BT tanks in the wheeled mode, added to the difficulties in technology associated with producing a tank which is able to travel on both wheels and tracks is required, it is suggested that future efforts should be directed towards the development of a less complex vehicle, running on tracks alone and employing coil sprung (Christie) suspension of the BT series.

This clearly bore fruit as following final approval of the drawings for the A-20 and the newly designated A-20G full-tracked machine by both the Military Council of the Red Army and by the USSR Defence Committee in February 1939, the Kharkov plant was ordered to proceed with the construction of prototypes of both machines.

Three months later these new tanks embarked on factory trials at the Kharkov works to be followed two months later, in July, by field tests. The rapidity of their construction was betrayed by some of the equipment used. For example, the A-32, as Kharkov now called the A-20, so designated because of its 32mm frontal armour, utilised a number of borrowed BT road wheels. Both machines were subjected

to extensive test drives at the end of which the chairman of the test commission nonetheless chose not to proffer a recommendation. On 23 September, and just 20 days after war in Europe had broken out, there was a major demonstration of all new armoured vehicles under development at the Kubinka Proving Grounds outside Moscow. Senior members of the Politburo were present, as was Pavlov and all the designers of the tanks on display, which in addition to the products of the Kharkov Zavod, included the KV and the very heavy SMK and T-100 prototypes from the Leningrad Works. Both the A-20 and A-32 were put through their paces, with the latter machine generating the most positive impression, especially when it charged head-on into a very large pine tree and knocked it over! It had been noted that the high power-to-weight ratio of the A-32, available by virtue of not having to be both wheel and track and thus carry the complex mechanism for that purpose, allowed for a weight growth in the amount of armour that could be carried. This was now recommended to be increased to 45mm.

On return to Kharkov, the A-32 was loaded up with weights to simulate the increase to 45mm of armour. The trials indicated that the chassis and suspension could cope with the heavier armour. This was grist to the mill for Koshkin and Morozov, as they were already engaged in the construction of two new tanks under the designation of A-34. Building on the A-32 with the heavier armour, this was not compromised by the need to be equipped for wheels and track operations, with both being designed to be wholly tracked from the outset. And strange as it may seem, the decision to build the A-34 had been taken 'in house' and without the sanction of any higher body.

Nonetheless, and in spite of the technological challenges raised by the production of these two new machines, in the same meeting of 21 October 1939, during which the decision was taken to disband the Tank Corps and give the tanks over to infantry support, it was agreed to re-equip the tank brigades with the A-34 medium tank. Notwithstanding the Germans and Russians professing to be on friendly terms – the non-aggression pact was barely two months old at this point – the underlying cynicism that governed the motivations of both parties was that it was viewed by them at best as a period of restraint before the inevitable armed confrontation that both knew would eventually come. For Stalin, what mattered was that this 'hiatus' be as long as possible so that the Soviet State and the Red Army could prepare itself for the coming war and this included acquiring the best possible military equipment for the armoured forces. Given the significance of this decision, it is worth quoting in full what the document sanctioning the selection of the A-34 stated:

> *Proceeding from the results of the demonstrations and tests of new tanks, armoured vehicles and prime movers, built in compliance with Defence Committee's Resolutions No. 198ss dated 7 July 1938 and No. 118ss dated 15 May 1939, the Defence Committee under the Council of People's Commissars of the USSR resolved:*
> 1 *To adopt for service with the Red Army: The T-32 fully tracked tank powered by the V-2 diesel engine, as developed by Plant No. 183 with the following design changes:*
> - *Main armour plates thickness increased to 45mm;*
> - *Crew vision improved;*
> - *The following armament installed:*
> i *The 7.62mm F-32 gun with co-axial 7.62mm machine gun.*
> ii *A 7.62mm radioman's machine gun;*
> iii *A 7.62mm fixed machine gun [author's note – a spare carried inside the tank]*
> iv *A 7.62mm anti-aircraft MG.*
> *And designate the tank A-34.*

Thus after a tortuous development which began in October 1937 with the issue of a specification for the BT-20 wheel-cum-track light tank – itself a development of the BT-IS tank of 1936 – we arrive in late 1939 with the official sanction for the production of a new, fully tracked medium tank that was to re-equip the Red Army and which came to be known subsequently as the T-34.

It had been adopted sight unseen, for the first A-34 was not rolled out of the Kharkov works until January 1940. Although it could

EVOLUTION OF THE MEDIUM TANK FROM THE A-20 TO THE T-34

	A-20	A-32	A-34/T-34	T-34 Mod 1940 (production model)
Year of prototype	May 1939	May 1939	January 1940	June 1940
Weight (metric tons)	14.6	18	19	26.3
Main gun	45mm	45mm	76.2mm	76.2mm
Max armour	22mm	25mm	30mm	45mm
Engine type	V-2 diesel	V-2 diesel	V-2 diesel	V-2 diesel
Drive configuration	Wheel-cum-track	Wheel-cum-track	Track	Track

be argued that this showed great faith in the design ability of Koshkin and his team, the decision has to be set against the backdrop of the new war that had broken out in Europe and the uneasy Non-Aggression Pact that the USSR had recently signed with the very state it saw as its most likely enemy in a near future war.

Testing times

Both prototypes of the A-34 had been rolled out by the end of February 1940. They were then, as part of the formal testing procedure, to undertake the standard field test of a 2,000km drive. This had to be completed by March when both machines were expected to be present in Moscow to be examined by government officials, including Stalin. Under normal circumstances this would have been carried out and completed around Kharkov as had those for the earlier A-20 and A-20G, but problems with the V-2 diesel engine in the first tank saw it break down after just 25 hours of use. Although replaced, neither machine had covered anything like the required 2,000km proving drive by 26 February. In the face of the approaching deadline in Moscow, which under no circumstances could be missed, the senior management at Zavod No. 183 decided that the only way to make up the required distance was to have both machines undertake a 'race' from Kharkov to Moscow. Koshkin was handed the poisoned chalice of organising it!

In the days that followed, Koshkin and his small team set down a detailed and very tight timetable that would allow both tanks to arrive in Moscow by the due date. The route chosen took the group of machines and their crews and support staff away from any centres of population and away from prying eyes because the two tanks were officially 'top secret'. Spare parts, food and other consumables were carried in two Voroshilovets fully tracked prime movers – one of which was fitted out for the tank crews to sleep in. What would under any normal circumstances have been a very challenging journey was made even more problematic by having to be carried out amid severe winter conditions. Snow, ice, extremely hard frosts at night and low temperatures in the day would test the two tanks to the limit. It would also be particularly hard on their crews as none of the machines carried any method for providing heat. The temperature inside the tanks would be nearly as low – many degrees below zero – as on the outside. It was under these conditions that Koshkin contracted a cold that was to lead to the pneumonia that would kill him before the summer of 1940 was out.

Nonetheless, and in spite of one of the tanks breaking down en route, both made it to the presentation for the morning of 18 March which was held within the Kremlin walls in the Ivanovskaya Square. Present were Stalin, Beria (a number of whose NKVD staff were

BELOW The second prototype of the A-34 tank. This was the immediate progenitor of the T-34 medium tank and seen on trials, also at Kubinka, in March 1940.
(Gennady Petrov)

sitting inside the two tanks alongside Koshkin and drivers from the Kharkov works), Molotov, Voroshilov and other functionaries from the Party and Pavlov. Koshkin's presentation was interrupted by his frequent coughing, to the displeasure it would seem of both Stalin and Beria, but with its completion both machines were then demonstrated to the general approval of most save for Pavlov and Kulik. It was noted by these two that there were numerous defects that would need to be addressed before the new machine could be committed to production. Whereas Pavlov had taken umbrage at the T-34 because it was not the type of improvement of the BT he had been looking for, in Kulik's case he seems to have just developed an antipathy to the design and was exercising his power as Deputy Commissar of Defence to stymy its further development (his comeuppance came later in 1941 when Stalin sacked him for incompetence). In this matter he was blocked by the intervention of V. Malyshev who, as People's Commissar for Heavy Machine Building, also became People's Commissar for the Tank Industry – a role in which he was to serve through to 1956.

The next stage in the validation of the two machines was the scrutiny they would be subjected to at the Kubinka testing grounds. The upshot of these, which assessed the coherence of the armour and put the machines through further driving trials, revealed the need for additional modifications to be carried out before the sanction could be given for the A-34 to be put into production. One of the T-34s was despatched to the northern province of Karelia where it was put through its paces in severe winter conditions (albeit it was not tested against the Finns). When, on 31 March 1940, the modified A-34 was made available for examination, the decision was taken to sanction the production of what was now called the T-34. This was formally confirmed with the issue of Protocol No. 848, which empowered the mass production of the new medium tank on two sites, of which the primary was Zavod No. 183 in Kharkov, and the other STZ in Stalingrad. Of note was the recommendation to enlarge the turret in order to enable crew to have more space, but to do so without compromising either the size of the turret ring or the slant angle of the

ABOVE As part of those same trials the A-34 is put through fire-extinguishing tests. The very early raised driver's hood can clearly be seen. The searchlight carried over the gun was deleted on the production version of the T-34. *(Gennady Petrov)*

ABOVE An early T-34 equipped with the short 76mm L-11 main gun. The boxes on the hull sides carry extra fuel. *(Gennady Petrov)*

BELOW Final tests of the T-34 prototype were carried out in Karelia where the Soviets had recently fought a war against the Finns. *(Gennady Petrov)*

amour on the turret. The two T-34s were returned to Kharkov in the same manner that they had travelled to Moscow – on their own tracks. But the wear incurred on a number of the elements of the motor, clutches and gearbox indicated that the new T-34 would not be able to meet the requirement to achieve the 3,000km range without breakdown required in the specification. Nor could they properly address other matters that needed attending to, of which failure to resolve would impact on the performance of the T-34 in the first two years of the conflict.

Although the management of the Kharkov factory now made plans for the production of 150 T-34s, this was circumvented by a resolution of 5 June 1940 issued by the Council of USSR People's Commissars and the Central Committee of the Communist Party on the production of T-34 tanks which required a far larger production order. It stated that:

> Taking account of the special importance attributed to the need to equip the Red Army with T-34 tanks . . . it was resolved that:
> 1 To empower People's Commissioner for medium Machine-Building Comrade Likhachev:
>
> a to supervise the production of 600 T-34 tanks in 1940, of which
> 500pcs. to be built by Komintern Plant No. 183, and
> 100pcs. to be built by the Stalingrad tractor Works (STZ);
> b to ensure supplies of engines in fulfilment of the 1940 programme of T-34 production, for which V-2 diesel production at Plant No. 75 should be enhanced, with 2,000 engines built before the end of 1940.

The NKO's decision to proceed with the full-scale production of the T-34 was just one element in a wider programme whose purpose was to replace obsolescent machines with newer designs. The T-40 amphibious light tank was to replace the T-37 and T-38, with the new T-50 replacing the T-26. Because the T-50 proved both expensive and problematic to produce, the programme was terminated after just 65 had been built. The BT series was to be replaced by the T-34, and T-28 and T-35 by the KV-1 and 2 heavy tanks. That the June 1940 resolution was not enacted upon came about as a consequence of a major change in the development of the T-34 tank. Had this come to pass it would have seen the machine that is the subject of this book relegated to a mere footnote, its place on the production line taken by what would have been the definitive design of the T-34/76 and one very different to that designed by Koshkin and Morozov.

BELOW A photograph set up to illustrate the development of the T-34 from its origins in the BT-7 on the left through the A-20, the T-34 1940 to the 76mm L-34 armed Model 1941. The latter variant first underwent tests in late 1940 and entered production in March 1941. *(Gennady Petrov)*

The T-34 that never was – but might have been

It was a bitter irony for the Russians that many of the limitations of the T-34 that they had identified even before it went into production in June 1940, were to have been addressed in a new variant which was to have superseded it. As such, the T-34 that is the subject of this book would have become an interim machine pending the introduction of the T-34(M), where the (M) stands for modernised. Even though this new design was on the cusp of entering production, the German invasion of June 1941 led to the immediate cancellation of the T-34(M) programme. It is just a fascinating 'might have been' in the T-34 story but worth telling because it illustrates that the Russians clearly understood the limitations of the T-34 from the outset and long before they were pointed out by others.

Although the primary function of the Non-Aggression Pact signed between Nazi Germany and Soviet Russia in August 1941 was the alleviation of tension and the maintenance of peace between the two powers, trade also flowed between them. Whereas Germany benefited from the supply of grain, oil and other raw materials, the USSR expressed a desire to acquire examples of the latest German military technology. In addition to military aircraft it also led to the purchase of two Panzer IIIG medium tanks in 1940. These machines were to have a profound impact on medium tank development in the USSR. Upon arrival in Russia, one of these tanks was despatched to NIIBT – the Scientific Institute for Armour Technology at Kubinka – and the other to Zavod No. 183 in Kharkov, where it was taken apart and examined in great detail. At Kubinka, the Pz.Kpfw III was employed in comparative trials with a prototype T-34, which, because of the numerous teething problems it was exhibiting at this time, was not held in great esteem. Comparison with the Panzer III came as somewhat of a revelation to the Soviets, drawing as it did attention to many weaknesses in the design of the T-34.

While there was no question that the T-34 was superior to the German tank in both armament and armour protection, in all other aspects the Mark III was better, with the sum of these elements prompting the development

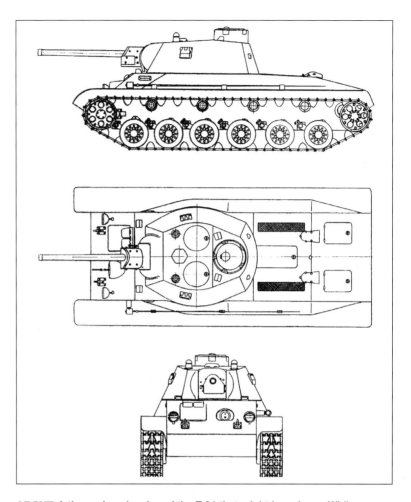

ABOVE A three-view drawing of the T-34 that might have been. While retaining aspects of the T-34 it can be seen how the turret is made bigger in order to accommodate three men, and the chassis has eschewed the Christie suspension in favour of a torsion bar system. The commander has also been given a cupola. *(Gennady Petrov)*

of a new version of the T-34 to accommodate them. Although the original designation was to be retained, what would have emerged was, to all intents and purposes, a new machine that, while retaining a superficial resemblance to its progenitor in the slope of the hull, was otherwise completely different. Indeed, the only element retained from the original T-34 was the shape of the hull glacis, which had the slope reduced to 52 degrees from the 60 degrees on the T-34. The driver's hatch and the gunner's position were swapped so that the former was now on the left, not the right of the glacis. Although the 76.2mm F-34 would continue to be the main armament, it was to be mounted in a cast mantlet, a feature not introduced on the standard T-34/76 until 1942. However, the new design would be heavier than the original T-34.

The most significant difference lay in the adoption of a torsion bar suspension in place of the Christie type inherited from the earlier BT series. Given the rapidity with which the T-34(M) was developed, it is likely that the decision was made to employ that already being utilised on the heavy KV tank, along with its road wheels. Also of note was that the T-34(M) was designed to be shorter than the T-34. This was because the engine – a more powerful version of the V-2 diesel developing an extra 100hp – was to be fitted transversely in the hull. This took up less space, enabling the fighting compartment to be enlarged. It was also to be connected to a new and far more effective gearbox that would have eight forward and two reverse speeds, thereby redressing substantially one of the primary weaknesses of the T-34. A larger turret ring was also to be fitted by widening the hull. The introduction of a larger, drop-forged, 45mm-thick, three-man turret made for a far more rational workload with the commander no longer having to double-up as the gun loader, something that we will see was to profoundly impact upon the efficacy of the T-34 in combat.

That role was now being taken by the addition of a third man/loader. This was also derived from that of the Pz.Kpfw III. Also fitted to the turret, in emulation of the German machine, was a cupola affording the tank commander a superior view of the battlefield either by employing vision blocks mounted on it or permitting him to survey it with his head outside, the latter allowing him the best possible situational awareness. The number of vision devices on the turret was also increased. Collectively these changes substantially improved what could be seen by the turret crew compared with the very limited vision of the two-man turret of the T-34. However, unlike on the German machine, the Russian design eschewed the hatches on either side of the turret thereby requiring the turret crew to have to enter and vacate the tank via the cupola.

The T-34(M) design was approved as early as January 1941, with the assembly of two prototypes beginning in March. On 5 May 1941 the Council of the People's Commissars of the USSR sanctioned the production of 2,800 T-34 tanks from Zavod No. 183 and STZ, as well as authorising the production of '500 improved T-34s in 1941 within the framework of the programme to be launched with this resolution'. Two prototypes were ordered to be ready by August. Such was the impetus behind this programme that Zavod No. 183 had produced three hulls of the T-34(M) by 17 April with five of the new turrets being manufactured by the Ilyich Metalworks at Mariupol by the beginning of May. The German invasion less than two months later brought this ambitious plan to a halt, it being superseded within days of the outbreak of war by another resolution from the presiding bodies in the Communist Party, setting forth a massive increase in the production of light, medium and heavy tanks already in production. Thus the T-34 in its unmodernised form became the medium tank of necessity.

Order, counter-order, disorder

With the German victory in the West and the surrender of France at the end of June 1940, the rationale that had underpinned Stalin's decision the previous August to opt

BELOW A wooden model of the T-34(M) as it was designated was all set to go into production in 1941 but the German invasion brought the whole programme to a halt. In this design many of the limitations of the T-34, and of which the Russians were all too aware even before the war began, had been eliminated. Much of the work that had gone into this design fed into the T-43 of late 1942. *(Gennady Petrov)*

for a Non-Aggression Pact with Nazi Germany tumbled down about his ears. Although it was regarded by either side – notwithstanding the official professions of friendship – as effectively an armed truce, for Stalin the great gain as he saw it was that it would perforce tie Germany down in the West into fighting a long and gruelling conflict that would serve to drain its resources, as had been the case in the First World War. In that hiatus, the Soviet Union could prepare and the Red Army rebuild and equip itself for the inevitable conflict with Nazi Germany. That assumption vanished with the French surrender in June 1940. Indeed, Stalin's apparent despair is certainly discernible in his response on hearing the news: 'Couldn't they put up any resistance at all? Now Hitler's going to beat our brains in!' At the time the 'they' included Great Britain, but with the rejection of the peace overtures from Hitler by that island nation's leaders, and its continuing resistance throughout the late summer and autumn of 1940, Stalin was somewhat mollified in his fears. He now convinced himself, to the degree that it became for him an *idée fixe*, that Hitler would not attack Russia while Britain remained undefeated. He did not believe that the Führer would repeat the great error of fighting the *Zweifrontenkrieg* that had condemned Germany to defeat in the First World War. This was a fallacy that Stalin would remain wedded to until the invasion began.

Although the poor performance of Soviet mechanised force in the Russo-Finnish War of 1939–40 had already prompted the beginning of a rethink of its organisation, it was the profound shock of the speed with which the Germans defeated France in the early summer of 1940 that provided the catalyst for a programme of massive change in the Red Army. Not only did it drive the realisation in the Red Army that the decision of October 1939 to abolish the tank corps had been a grave error, it also generated no little bitterness that the German achievement in Poland and the West had been realised by them embracing ideas on armoured warfare similar to those that had been propounded, traduced and abandoned by the Red Army some years before. All now became directed towards managing a huge reversal of the October 1939 decision which came with the order on 6 July 1940 for the NKO to create nine new mechanised corps. This was to be followed in February and March of 1941 by a further 20. To be overseen by the new defence minister S.K. Timoshenko, this massive programme – although embraced and executed with haste, owing to the changed international scene – was nonetheless only expected to be complete by the summer of 1942 at the earliest.

There is no question that had it been realised in the manner envisaged it would have seen the creation of a new Red Army based upon an immensely powerful new armoured force. Each of the new mechanised corps was larger than its predecessor and on paper was to be comprised of two tank divisions with a total of 126 heavy KV tanks and 420 T-34 medium tanks and one mechanised division with 275 light tanks and 49 armoured cars. Had this programme come to fruition as planned, by the summer of 1942 there should have been no fewer than 12,000 T-34s on strength. That such a figure could have been achieved, however, is unlikely, even if at the end of October 1939 Voroshilov had told Stalin that Soviet industry was expected to be producing 3,600 T-34s per annum by 1941. The People's Commissariat for Defence (NKO) was more realistic in estimating that it would take until 1943 before industry could deliver anywhere near the totals required to equip all 29 mechanised corps.

As the war clouds gathered in the East during the first part of 1941, all was being done to expedite the programme. While all 29 mechanised corps had been formed by June, only the first nine set up in July of the previous year were near readiness and many of those were very far from such. As we shall see, according to David Glantz, by the time of the launch of Barbarossa: 'Most of these corps were still woefully understrength in manpower, equipment and logistical support, and the officers and men who manned them were largely untrained.'

And of the nine mechanised corps set up in the first tranche and which were stationed closest to the western border of the Soviet Union in June, only 1,475 T-34s out of the theoretical total of 3,780 which should have been on strength, were actually available.

Chapter Two

T-34 at war 1941–45

Although most of the T-34s in service with the Red Army were lost following the German invasion in June 1941, its revolutionary design profoundly affected the course of the conflict. Subsequently built in vast numbers, the T-34 became the mainstay of the Soviet tank formations and a primary weapon in helping to secure the defeat of Nazi Germany four years later.

OPPOSITE This posed propaganda image permits us a very clear view of a T-34 Model 1941 with a cast turret. This type was in production from the autumn of 1941 through to the spring of 1942 – its identifying features being the twin periscopes and the ventilator cover mounted between them. It carries a number of the early long box-type external fuel tanks. The hull has twin periscopes above the driver's hatch that marks it as an earlier production machine, but it also mounts the newer towing hooks that appeared in the summer of 1942. *(Gennady Petrov)*

Operation Barbarossa – the German invasion of the Soviet Union – began in the early hours of 22 June 1941. According to Adolf Hitler, the German Führer, 'We have only to kick in the front door and the whole rotten structure will come tumbling down.'

At the time, this low opinion of the effectiveness of the Soviet Union held by the German leader was one shared by the higher echelons of the Wehrmacht, as was the expectation that the time needed to defeat the 'colossus in the East' would only be a little more than it had taken to defeat France the previous year. Despite the rash of indicators and intelligence summaries detailing German intent reaching the Kremlin in the months prior to the invasion, in appropriating to himself the sole right to determine the veracity of these reports and their implications, Stalin wilfully chose to ignore them. In consequence, the order was fed down to all command levels of Red Army formations deployed in western Russia that they were to do nothing that could indicate to the Germans any hostile Soviet intent by raising the readiness of these same units. Suffice it to say when the invasion was launched by the Wehrmacht and its allies on a front stretching from the far north of Russia down to the Black Sea, it realised strategic and tactical surprise of such a magnitude that within hours of its opening, the Red Army was reeling and being thrown back at those points where the German assault was concentrated. Within days, the four Panzer Groups of Army Groups North, Centre and South, had cut swathes through the border defences. Supported by the Luftwaffe, deporting itself in skies swept clear of the Red Air Force, the Germans were already pushing deeply into the Russian hinterland. On 27 June, just five days after the start of the invasion, the leading elements of Panzer Groups II and III had linked up to the east of the city of Minsk, forming the first of the great encirclements that was to be such a feature of the opening six months of German operations in the Russian Campaign.

The cutting edge of Operation Barbarossa – the Panzers

The T-34 (and KV) shock

By 27 June, the T-34 had already been blooded in combat alongside the heavy KV-1 and KV-2 tanks. However, determining just how many T-34s were actually in service at the time of the German invasion has proven to be a problematic task. The most reliable figures put the number at 967, although the monthly strength return for 1 June 1941 lists the lower number of 832 T-34s deployed in the four

RIGHT Of the 3,538 tanks with which the Germans invaded the Soviet Union on 22 June 1941, 966 were of the Pz.Kpfw III medium tank, the primary AFV of the Panzerwaffe at the time. However, only 707 of these mounted the 50mm L/42 gun with the remainder still equipped with same ineffectual 37mm weapon as used by the German anti-tank units. The appearance of the T-34 rendered this machine obsolete overnight. *(Author)*

RIGHT Although from 1943 onward the Pz.Kpfw IV would become numerically the most important tank in the Panzerwaffe, in June 1941 it was still serving in its designed role of 'heavy support tank' to its stablemate, the Panzer III. Only 439 were in service for the invasion. *(Author)*

main military districts in closest proximity to the Soviet western border. Not all T-34s available to the Red Army would have found their way to these front-line units – others being held back for development work, training and service with other formations. Nor was the distribution of T-34s operating with these military districts equal. There were many more in service with the Kiev Military District, which was deployed to the south of the Pripet Marshes covering western Ukraine, as it was here that Stalin believed Hitler would concentrate his primary assault. The number of T-34s in service with this

RIGHT The Pz.Kpfw 38(t) was a light tank mounting a 37mm gun and was roughly the equivalent of the BT series. It armour was thin and as with the two other light tanks in service simply no longer adequate to the task they had been set. Nonetheless, there were 625 in service with five of the panzer divisions committed to the invasion. *(Author)*

LEFT The Panzer II was employed as a reconnaissance machine and was in service with all panzer divisions committed to the invasion with 743 on strength. Its armour was thin and its armament was just a 20mm cannon and a machine gun. *(Author)*

ABOVE LEFT The Panzer 35(t) was the second Czech design integrated into the Panzerwaffe when that country was incorporated into the Reich. It can clearly be seen from the image that it was constructed employing rivets – a retrograde feature even in 1941. Only 155 were used in the invasion and they served with just the 6th Panzer Division which was unfortunate in being one of the first to encounter the T-34 and KV-1 with disastrous results on 23 June. *(Author)*

ABOVE There were 272 Sturmgeschutz IIIs serving in 18 StuG units at the time of the invasion. Equipped with the same 75mm L/24 cannon as the Panzer IV, the machine had been designed to support the infantry but very quickly was employed in an anti-tank role in which it proved to be highly effective. *(Author)*

CENTRE LEFT Most of the T-34s lost in the opening weeks of the German invasion were of the 1940 model with the L-11 gun. *(Author)*

LEFT This T-34 1940 was driven into a ditch and abandoned. It can be seen how large the turret hatch was and how difficult it was to open because of size and weight. *(Author)*

LEFT A rare colour image taken by a German PK (Propagandakompanie) cameraman showing three T-34s – one of which in the centre is an early Model 1941. Many Soviet tanks were lost in this way as when under air attack they pulled off the road and into the very many watercourses that littered the landscape. Being covered in vegetation these were difficult to spot before being driven into in the heat of combat. These T-34s belong to a unit of the Kiev Military District. *(Copyright unknown)*

and the other four military districts deployed to cover the USSR's western frontier are detailed in the table below.

The German estimate of the size of the Soviet tank park prior to the invasion was that it numbered some 10,000 machines. This was a

Military district	Mechanised Corps	Divisions in Mechanised Corps	Total no. of tanks	T-34 and KV
Baltic Special	3	2, 5, 84	651	110
	12	*	749	N/A
Western Special	6	4 ,7, 29	1,132	452
	11	29, 33, 204	414	20
	13	*	282	N/A
	14	*	518	N/A
	17	*	63	N/A
	20	*	94	N/A
Kiev Special	4	8, 32, 81	979	414
	8	12, 34, 7	899	171
	9	*	298	N/A
	15	10, 37, 212	749	136
	16	15, 39, 240	482	76
	19	*	453	N/A
	22	19, 41, 215	712	31
	24	*	222	N/A
	5	13, 17, 109	1,070	17
Odessa	2	11, 15, 16	*	N/A
	18	*	282	N/A
			8,570	1,590

* = *Identity of divisions uncertain.*

ABOVE The roads down which the Soviet armoured units advanced were littered with abandoned T-34s in fairly good condition that had quite simply broken down or run out of fuel. *(Author)*

gross underestimate. In actuality the figure was 25,932, albeit with by far and away the majority being of the light T-26 and BT series. Even in the face of this massive numerical superiority, the much smaller German tank force numbering some 3,505 tanks and 250 assault guns helped bring about the loss of no fewer than 11,000 of the totality of the Soviet tank force by the middle of July. This included nearly all of the T-34s and KVs in service with the military districts closest to the western border. Such was the level of loss (not always destruction) that on 15 July 1941 STAVKA (the Soviet High Command) disbanded what remained of its Mechanised Corps. It may be splitting hairs to distinguish between tanks lost and tanks destroyed, but it is apparent from German combat reports that these colossal Soviet tank losses were not wholly attributable to their handiwork. In seeking to offer even a limited explanation for its occurrence, it needs to be recognised that there was a litany of factors at work to account for the almost total ineptitude exhibited by the Soviet military at this time. This is especially so when we seek to account for Soviet tank losses, as many of the reasons that led to the loss of the huge number of light tanks also provides an explanation for those of the T-34s (and KV tanks).

As we have seen earlier, the Red Army was in the midst of a massive restructuring and modernisation programme that was not expected to be completed before the following year at the very earliest. Although under Defence Minister Marshal S. Timoshenko the whole process was being expedited, it was so complex that the German invasion came at exactly the moment when the Red Army was least prepared to face it. David Glantz, who has written many detailed books on the Red Army, summed up the totally unready state of the Soviet tank forces:

On 22 June 1941, the Soviet mechanised corps in the border military districts and throughout the entire armed forces

RIGHT Within a month of the start of the invasion, many of the over 900 T-34s in service with the Mechanised Corps stationed near the frontier had been lost. *(Author)*

structure were not combat ready. They were understrength by 25 per cent in enlisted personnel and short even of greater numbers of command cadre and NCOs. Training of these personnel was poor, especially in the corps that had been formed in 1941. Corps' equipment strength, in particular tanks, averaged 53 per cent, and most tanks were older models that required repair or reconditioning. Compounding these problems, the corps was mal-deployed and lacked clear missions. The Soviet's immense mechanised forces did possess considerable combat potential. That potential, however, would take months or perhaps a full year to realise. As subsequent combat and the reports of its command personnel vividly indicated, the corps were simply not prepared to fight the world's most experienced army in the summer of 1941.

In practice, what this meant was that those T-34s and KVs already in service and which could have made a profound impact in blunting the German invasion, were quite simply frittered away, their technical superiority neutered by the many failings within the Red Army. Those few reports offered below are but exemplars of what was happening among most of the Red Army Mechanised Corps in these opening days of the war. It is these factors that account for the vast loss of tanks in the Mechanised Corps of the Red Army.

Major General D.I. Riabyshev, the commander of the 8th Mechanised Corps which was deployed to the south of the Pripet Marshes, noted that even while his command had 100 T-34s on strength – some 320 fewer than the authorised establishment – the majority of their drivers had received just three to five hours' driver training! Furthermore, having been recently formed, the Mechanised Corps as a whole had not been able to take part in hardly any 'tactical exercises' and had not been tested either in the matter of March training or in actions in the principal types of combat. In consequence, when in the four days after the invasion the corps embarked on 'super-forced' marches, some 40–50% of combat vehicles broke down for 'technical reasons'. Such experiences were hardly unique to this formation.

Logistical support for the new and older machines was in many cases extremely poorly organised with POL (petrol, oil, lubricants) supplies in some cases non-existent. Where tanks moved out with full fuel tanks, once used up there was no more available for replenishment, so their crews simply had to abandon them. Ammunition was sometimes or often lacking; tanks in some units went into battle without any. This can be illustrated by reference to the example of the 6th Mechanised Corps, 7th Tank Division operating in the Bialystock Bend. In a report penned some weeks after the invasion by its then commander Major General B.S. Vasil'evich, it was stated that at the outbreak of hostilities his formation had on strength 348 tanks of which 51 were KV 'heavies' and 150 T-34s. He was candid in stating that the division's weakness lay in its supply echelon. At the outset of its march to battle it carried only one to one-and-a-half combat loads of ammunition for the 76mm guns, none of which was armour-piercing, three refuels of petrol/gasoline and a single fill of diesel fuel. The receipt of a battery of confusing and contradictory orders from higher authority saw the direction of march changed three times in the first two days of the war. In consequence, the limited supply of fuel ran out and the whole division became immobilised south of Grodno.

Even where T-34s broke down there was a major lack of recovery machines and of the two types, namely the S-60 and S-65 Stalinets agricultural tractors that were in service, neither was capable of recovering the medium T-34 or

ABOVE Where possible, driveable T-34s were recovered by German crews as booty. This particular example has lost the third road wheel on its nearside. At this early period of the invasion the T-34 is still the object of great interest to all who see it. *(Author)*

ABOVE A T-34 that was a definite combat casualty. The loss of the second road wheel suggests that it was penetrated in that position possibly by an 88mm shell, which caused the ammunition stored in the floor containers to explode. The subsequent blast then lifted the turret off the hull. It looks as though the T-34 and BT-7 were moving in tandem at the moment the former was hit. It then ploughed into the bridge and the BT-7 then rammed into its rear. Whereas the crew of the BT-7 probably escaped, it is unlikely that those of the T-34 did. *(Author)*

KV-1 and 2 heavy tanks. The upshot of the latter observation was that if these machines broke down, they too had to be abandoned. This was the case with the 4th and 6th Mechanised Corps which were well equipped with both new tanks, with 313 and 238 T-34s on strength respectively. The bulk of these were lost even before meeting the Germans, owing to breakdowns on the route march to the front. There were constant changes of direction for units as their commanders responded to the rash of orders issued by higher authorities. In many cases these were out of date by the time they were received due to the rapidity of the German advance. Major General N.M. Shestopalov, the commander of the 12th Mechanised Corps deployed in Lithuania at the time of the invasion, noted that units belonging to the North-west Front had been ordered into combat in a piecemeal fashion by the front commander himself. Shestopalov later wrote that:

From the first days of combat, we began to receive operational orders or combat instructions two or three times daily which contradicted one another. As a result the forces 'twitched' in vain and this situation prevented the capability of using the forces and weaponry expediently to fulfil orders and deprived us of the ability to employ large formations.

RIGHT Another definite combat loss. On this T-34 Model 1940 the whole of the offside suspension has been demolished and smoke is billowing from the engine deck. *(Author)*

The chaotic conditions engendered among Soviet units profoundly reduced their combat effectiveness and was grist to the mill for the experienced German forces who were able to exploit such disorder, notwithstanding their need to deal with the unexpected appearance of the T-34 and KV tanks. A very informative summation was proffered to the Front Military Council on 3 July by Major General Morgunov, chief of the South-western Front Armoured Forces in which he details some of the many reasons for the massive tank losses incurred by that date:

The absence of evacuation means and reserve equipment for the KVs and T-34s; the presence of factory defects; the lack of familiarity with the tanks; insufficiently trained personnel; weak antitank reconnaissance of the enemy; systematic bombing on the march, in concentration areas and during attacks; extensive manoeuvring over 800 to 900 kilometres without aviation cover and artillery coordination over almost prohibitive (for tanks) forested-swampy terrain; strong opposition by a predominant enemy; and the absence of armour–piercing shell for KVs and T-34; has led to huge mechanised corps losses and lack of combat readiness on the part of those which remain.

It is hardly surprising therefore that what unfolded in the opening months of Barbarossa was an unprecedented catastrophe for the Red Army.

That there is a paucity of accounts from Russian sources dealing with the operation of the T-34 against the invader should not come as a surprise. Indeed, this hiatus extends until October 1941. For the initial period of Barbarossa we have therefore to fall back on German combat reports of their encounters with the T-34. The appearance of these new machines of which the German Army knew nothing, caused consternation among the troops in the field as they attempted to wrestle with these armoured 'monsters'. However, the 'knew nothing' may be in need of a degree of qualification. It may well have been that the Fremde Heere Ost – that section of German Military Intelligence that dealt with the Russian Army – had indeed got wind of these new Russian tank types. But what was known of them, beyond vague references, was not of sufficient substance to have enabled the German invasion forces to have been prepared for their encounters with them.

This was especially so for those whose task it was to service the weaponry that was to defeat enemy tanks. Even during the French Campaign the previous year, the Army's standard anti-tank gun, the 37mm PaK 35/36, had been found wanting. Although the 50mm PaK 38 was available in some numbers by the time of Barbarossa, with some 800 in service, even this larger-calibre weapon had great difficulty coping with these new Russian types. The reason for this was that the Germans had absolutely no expectation that they would encounter tanks like the T-34 and the KV-1 and 2. It was a salutary experience for them to be confronted with technologically superior machines that were virtually immune to the then standard German anti-tank weaponry. The T-26 and BT series of light tanks were known quantities as they had been encountered before and had proven vulnerable to the 37mm PaK 35/36 in the Spanish Civil War. The T-28 medium and the heavy T-35 were also known to be in service, albeit in very much smaller numbers. Their armour, although slightly thicker than that of the light tanks, would still prove vulnerable to the two main German

BELOW One for the family album! Two German soldiers pose for a photo on an abandoned T-34 Model 1940. In the bottom right foreground is one of the early external fuel containers carried by the tank which must have come adrift when it was hit. *(Author)*

ABOVE An abandoned T-34 Model 1941 lies next to the wrecked hull of a BT tank. The cause of the T-34's demise seems to have been that it lost its offside drive wheel. *(Author)*

anti-tank guns. This was also the case with the armament of the Panzer III, the standard medium tank of the *Panzerwaffe* and the most numerous German tank committed to Barbarossa. It mounted either a 37mm or a short 50mm weapon. In essence, these were the same anti-tank guns adapted for use in this panzer, although the latter weapon had a shorter barrel length than that of the PaK 38 and thus possessed a lower muzzle velocity. While the armament of the 'heavy' Panzer IV was of a larger calibre at 75mm, it had a low muzzle velocity and so had the same problems dealing with these new Russian types.

Thus the appearance of the T-34/KV-1/KV-2 truly did come as 'a bolt from the blue', invalidating from the outset the German presumption that the qualitative superiority of their panzer divisions would be more than sufficient to see off the superior numbers in the Soviet tank arm. Not only did they discover that they had profoundly underestimated the real size of the Red Army tank park, but also that these new medium and heavy tanks of which they knew nothing, were far superior to their best. It was a profoundly humbling experience and it impacted on the morale of German tankers. To discover that this new enemy, whom Nazi racist ideology was to deride and denigrate as *Untermenschen* (subhuman), was now fielding in the T-34, a medium tank superior to their Panzer Mark III in firepower, armour protection and manoeuvrability – the three

fundamentals of effective tank design. One German company commander remarked: 'This was a shocking recognition to the German tank and tank destroyer units, and our knees were weak for a time.'

The first encounter with the T-34 occurred on 23 June when tanks of the 7th Panzer Division belonging to Army Group Centre met a number belonging to the Soviet 5th Tank Division of 3rd Mechanised Corps in Lithuania en route eastward to Minsk.

Half a dozen anti-tank guns fire shells at him (a T-34) which sound like a drumroll. But he drives staunchly through our line like an impregnable prehistoric monster. . . . It is remarkable that Leutnant Steup's tank made hits on a T-34, once at about 20 metres and four times at 50 metres, with Panzergranate 40 [calibre 50mm – tank was a Mark III] without any noticeable effect.

It is worth noting that the contextual reference to *Panzergranate 40* is clearly one of frustration. This was a type of armour-piercing ammunition that formed part of the complement of shell types carried by the Panzers III and IV and was reserved for 'special targets' where a higher than normal muzzle velocity was required to defeat the enemy. That even this ammunition could not defeat the T-34, even at the very close range described in the account, sums up the measure of frustration and indeed impotence initially experienced by German tankers trying to defeat this new Soviet machine (and which would continue well into 1942).

In such circumstances, the Germans fell back on expedient methods – the hasty deployment of heavy artillery pieces firing over open sights and the 88mm Flak gun consequently became one of the primary fallback weapons in the event of attacks on German formations by T-34s and KVs. For panzer crews the means to deal with the T-34 (and KV) was arrived at by trial and error – the error being in the number of tanks lost to the T-34's 76.2mm gun. That described in a report from the 25th Panzer Brigade on 8 July describes the very dangerous nature of the undertaking: 'German tanks have a small chance of success if they can outmanoeuvre

Russian tanks and attack them diagonally from the rear or on the running gear, along with hits on the turret race.'

It then recommends that tank crews adopt a method being employed perforce by the infantry, of destroying them in close combat – on foot! This was something that infantry formations, possessing only 37mm PaK 36s for defence against tanks, had found out for themselves on the long march to the East. Seeing their shells bounce off the armour of T-34s and KVs saw them attribute to their anti-tank guns the highly pejorative label of *Die Armee Türklopfer* – 'the Army's Door Knocker'. In the face of the impotence of their anti-tank guns when confronting these machines, the infantry had to resort to improvisation, as is demonstrated in the following combat report. What is described here became the standard means for infantry to destroy these Soviet medium and heavy machines. In this case they employed engineering equipment.

> We had close combat and used material from the engineers, mostly mines, and all soldiers were trained in anti-tank use of these mines. In all rifle companies we had groups of three men who worked together. One man handled security and a second operated on the blind side of the tank with a mine and tried either to place the mine on the tank's rear hatch or use a hand grenade bundle (held together with wire) thrown over the tank barrel. A third method involved using a shaped charge, which was magnetic, emplaced against the tank. But the Russians countered this by placing concrete on the armour plate so the mines would not stick.

Other expedients adopted included turning captured examples of the Soviet 7.62cm M1936 or M1939 field gun, which was captured in vast numbers. These weapons were of the same calibre as the main armament of the T-34 and KV-1.

To counter the impact of these new Russian machines, within a month of the invasion, the OKH (the High Command of the Army) had seen fit to issue D343 *Merkblatt für die Berkämpfung der Russischen Panzerkampfwagen* – a leaflet obviously compiled from the collective accounts

ABOVE The frequency of images taken by the Germans (many soldiers carried their own personal Leica cameras throughout the war and in all theatres of conflict) of abandoned T-34s attest to the view that the vast majority of those lost by the Red Army in the first two months' campaigning were the consequence of mechanical breakdown and not German action. That the rear deck was raised suggests that the Russian crew attempted to see if it was possible to repair their charge before the decision was taken to abandon it. *(Author)*

BELOW Soldiers inspect a T-34 that has come to grief by driving over and then grounding itself on a German 105mm FH. This weapon was one of those which were employed to defeat the T-34 in the early days of the conflict. *(Author)*

ABOVE Until Plant No. 183 was evacuated in September/October it continued building and delivering new T-34 Model 1941s. It was only after that date that problems arose as the STZ was the only factory turning out new-build T-34s for a few months. *(RGAKFD Krasnogorsk via Stavka)*

A crucial role played in the destruction of Soviet armour – and not just of the T-34 or KVs – was by the Luftwaffe. By the time of Barbarossa the co-operation between German aircraft units – especially the *Stukagruppen* – had been raised to a fine art. A new technique introduced with the Russian Campaign was the attachment of a *Panzer Verbindungsoffizier* (tank liaison officer) to the armoured units leading the advance. These were Stuka pilots who, operating on a rotating basis, were in direct radio communication with the dive-bombers who could be directed more rapidly and efficiently on enemy targets when they presented themselves. Many Red Army columns were attacked as they moved towards the front line, resulting in the destruction of many tanks including T-34s. Such was the rate of loss among the T-34s in service on 22 June that only 235 were operational in early August, while another 116 were available in armoured units in the process of mobilisation. These numbers were to decline even further before the end of the year.

The Battle of Mtsensk

The paucity of accounts of the T-34 in action in Soviet sources lasts the better part of June through to November 1941. Indeed, the very first detailed account comes from October 1941 during the course of Operation Typhoon –

and advice to be found in the many combat reports fed back from the front about how these Russian tanks had been dealt with in the field. It became the first of many that would be issued on this subject. All of these methods would need to be employed until new weapons appeared that could deal with these Russian tanks.

RIGHT The 57mm Zis-4 anti-tank gun was first tested on a T-34 before the German invasion. It could penetrate a 70mm armour plate angled at 30 degrees. Although it was committed to production at Plant No. 183 and the STZ, only 42 were built. Production was terminated because of difficulties with the gun. It was given the name 'Exterminator'. Limited production was resumed in mid-1943. *(Nik Cornish at www.Stavka.org.uk)*

ABOVE Two shattered T-34 Model 1941s lay abandoned on a roadside in late 1941. The ground around them is littered with various parts of the tanks, such as road wheels. *(Gennady Petrov)*

the German offensive to capture Moscow – but it also figures in German accounts, primarily in the memoirs of General Heinz Guderian who seems to have employed the superiority of the T-34 as an alibi for the temporary German setback. As such it is necessary to be mindful that both the Soviets and the Germans sought to derive propagandistic value from what was at most a localised encounter. Nonetheless, the ramifications of the event were to be of major significance to the emergence of the specification for a new tank to replace the Panzer III.

Guderian had command of Panzer Group III which was the southernmost of the three armoured formations involved in the drive on Moscow. Following the capture of the city of Orel by the 24th Panzer Korps on 6 October, 4th Panzer Division began an advance on Mtsensk, which lay some 50km to the north-west, during the course of which it was given an extremely 'bloody nose' by an unexpected Soviet counter-stroke. Of interest is that the Soviet tanks were under the command of officers who would acquire a reputation and gain kudos in the Red Army as the war progressed. In charge of the 1st Tank Battalion of General D.D Lelyushenko's 1st Guards Rifle Corps was Colonel Mikhail Katukov (who would go on to command 1st Tank Army at Kursk). Having been ordered to take his tanks to the Mtsensk area, he had despatched two patrols on 4 October. On the following day Katukov's tanks attacked a German tank column. In an after-battle report, the 4th Panzer Division noted that:

The Russians employed their heavy tanks en masse for the very first time. In several

49

engagements it came to very hard tank battles, because the Russian tanks no longer let themselves be driven off by artillery fire.... For the first time in the East, in these battles the absolute superiority of the Russian 26-ton [T-34] and 52-ton [KV-1] tanks over our Pz.Kpfw III and IV was felt.

Clearly this was not the first time that the superiority of the T-34 had been felt. This had been known since the second day of the invasion, but what is meant here is not only were they encountered in more than ones and twos, but also that the Russian tankers had given thought as to how best to employ their machines and their superior firepower and how these concerns now informed their tactics.

The Russian tanks usually formed in a half circle, open fire with their 7.62cm guns and our panzers already at a range of 1000 metres and deliver enormous penetration energy with high accuracy. . . . Our panzers are already knocked out at a range of several hundred metres. Many times our panzers were split open or the complete commander's cupola of the Pz.Kpfw. III and IV flew off from one frontal hit . . . the accuracy and penetration ability of the Russian 7.62 cm tank guns are high.

Elsewhere in the divisional daybook it was reported that, 'The Russian was very skilled at directing his tanks, pulling back often, only to appear again in a flanking attack. In the course of the afternoon his heavy models inflicted heavy losses.'

The author of the report, Freiherr von Langermann, the divisional commander, also pointed out the ability of the T-34 to travel over rough and muddy ground at a higher speed than the German tanks, drawing attention to their lower ground pressure by virtue of having wider tracks better able to distribute the greater weight of the T-34. This was significant given the extremely muddy conditions through which the Germans were trying to advance. It would become even more difficult with the coming of the deep snow of winter. Von Langermann appended a whole raft of recommendations, including new weaponry, that it would be necessary for the Germans to acquire in order to master these Soviet machines.

The consequence for the advance of 4th Panzer Division on Mtsensk was that it was initially held up, then thrown back in retreat. The Germans acknowledged that the Russian force was smaller than their own – counting 45 Russian tanks of different types compared to 56 of their own. By the end of 6 October, 17 Soviet tanks had been destroyed and 10 German. The Russians held the field. The following day the Germans resumed their attack only to be pushed back once again. By 9 October, 4th Panzer Division was down to just 30 operational tanks with little to show for the losses expended.

RIGHT A T-34 Model 1941 bedecked in snow camouflage and followed by a BT tank moves through Moscow prior to the launch of the counter-offensive on 5 December. *(Gennady Petrov)*

Once more an attempt was made to resume the advance but the 'superior weaponry of the Russian tanks' again took its toll with four panzers being destroyed, while some were damaged and there were additional losses of supporting equipment. No Russian tanks were destroyed. The runes were not wrong as Guderian himself observed when he stated of the Russians after this battle, 'They were learning.' Much of von Langermann's document would be employed by him during the visit of a special *Panzerkommission* on 18 November 1941. From the ensuing discussion was to emerge the specification that would result in the Panzer III's replacement – the Panther. (The commission will be considered in much greater detail in a forthcoming Haynes manual on the Panther.)

With launch of the Russian counter-offensive on 5 December 1941, German tank strength in Russia was a shadow of that with which it had begun Barbarossa less than seven months before. It was not the case, however, that the number of tanks Zhukov had access to for this operation was excessive. Indeed, just 770 tanks were available to launch the counter-offensive and T-34s were not available in large numbers. And even though regular production of the T-34 by this juncture centred on just one factory – the STZ in Stalingrad – it was still the case that the T-34 had emerged as the most effective tank on the Eastern Front. The following table details the output of T-34s from three factories producing them from the time of the German invasion through to December 1941. Zavod No. 183 at Kharkov ceased production in November due to its evacuation and relocation eastwards to Nizhne Tagil. Thereafter STZ became the most important producer. Plant 112 at Gorki built only a small number by the year's end.

RIGHT Zhukov, who oversaw the counter-offensive, later stated that he did not have enough tanks to defeat the German Army Group Centre. Nor were his efforts helped by Stalin diluting what few tank units were available by expanding the Soviet offensive to embrace more or less the whole of the front. Where the T-34 was committed it made a great impression – in this case T-34s of the 1st Guards Tank Brigade in March 1942. *(Gennady Petrov)*

Plant	112	183	STZ	Total
June	–	170	86	256
July	–	209	93	302
August	–	266	155	421
September	–	228	165	393
October	20	41	124	185
November	58	–	200	258
December	83	25	219	327
	161	939	1,042	2,142

LEFT Stalin's decision to stay in Moscow even as the rest of the government was moved to Saratov for safety's sake was a major morale booster for Muscovites. Much was made also of the decision to hold the parade in Red Square that celebrated the October Revolution of 1917 even as the 'enemy was at the gates' of Moscow. In a famous political poster produced for the event, Stalin addresses his troops over the large image of a T-34 with a fainter backdrop being provided by another T-34 and a KV-1 tank. *(Copyright unknown)*

EVACUATION OF THE FACTORIES AND T-34 PRODUCTION

On 30 June 1941, all matters pertaining to the conduct of the war were centralised in a newly constituted body called the *Gosudarstvenney Komitet Oborony* or State Defence Committee (GKO) under the direct command of Stalin. It assumed absolute control and authority over the state, the Communist Party, the military and the Soviet populace. This body in turn ordered the creation of new people's commissariats that were to oversee specific areas of production in all areas of the economy as part of transformation to service the needs of the war. All extant plans were suspended and replaced by others already prepared for such an eventuality. On 4 July, the chairman of *Gosplan* (the State Planning Commission), Nicholas Vosnesensky, was tasked by the GKO with putting together a far-reaching, detailed plan for what was to become a 'second line of industrial defence' to be created in the east by organising 'a coherent productive combination between the industries already existing in the east and those to be transplanted there'.

Such was the bureaucratic backdrop to the massive programme instigated for the transplantation of hundreds of industrial concerns vital to the ability of the Soviet Union to continue the war. A specially created Evacuation Council was to organise the dismantling and entrainment of all moveable property before they could be captured by the advancing German forces. The corollary to the evacuation had been set down in a directive of 29 June, in which Stalin demanded that where this could not be carried out, everything was to be destroyed – the famous 'scorched earth' policy in which the invader was to find nothing that could be of service to his war effort. Between July and November 1941, and in accordance with the schedule set down by the Evacuation Council, 1,523 industrial enterprises were transported eastwards, of which 1,360 were armaments works. Of these 226 were moved to the Volga area, 667 to the Urals, 244 to western Siberia, 78 to eastern Siberia and 308 to Kazakhstan and Central Asia.

BELOW This map illustrates the evacuation of the major tank factories in the late summer/autumn of 1941 from their sites in western Russia to safer locations to the east of the Ural mountains.

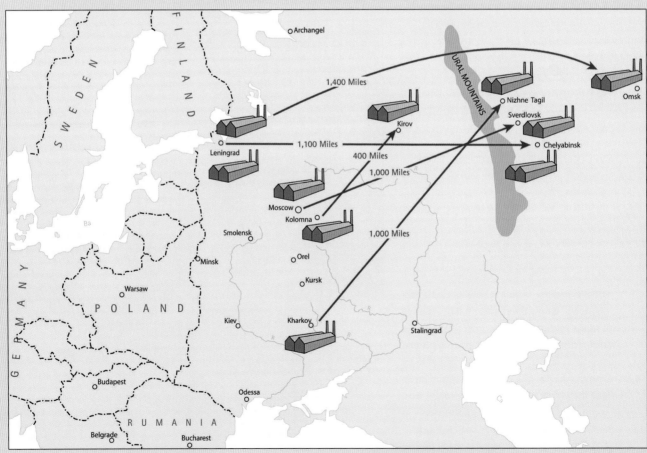

ABOVE May 1942 saw the launch of the disastrous Kharkov offensive by the Red Army. Carefully husbanded T-34s were committed to a badly prepared operation to recapture the USSR's fourth largest city. A complete misreading of enemy strength in the area led to the defeat of the offensive by the end of May with the loss of many hundreds of T-34s and other tanks. *(Gennady Petrov)*

The first enterprise involved with aspects of T-34 production was the Mariupol armoured plate mill which had already produced turrets for the T-34(M), the management being given its marching orders as early as 2 July. As the major plant for the production of armour plate, its security was absolutely vital. However, the primary production centre for the T-34 at Kharkov did not get its instruction to evacuate until over two months later when Zavod No. 183 manager Yu. E. Maksarev received State decision No. 667/GKO to do so on 12 September. It must be assumed that the issue of this order had been left so late in order to maximise production of the T-34 on this site to the very last possible moment. Just seven days elapsed between Maksarev curtailing production and ordering that all machine tools, spare parts, incomplete tanks, industrial workers and their families, the staff of the Design Bureau and all technical documentation pertaining to the T-34 be made ready to move, with the first train departing on 19 September. Over the next month, with NKVD guards overseeing the process, a further 40 trains were loaded up with each making its slow, snaking journey eastward beyond the Urals. The very last one pulled out of Kharkov just two days before troops of the German 6th Army began battling Soviet 21st and 38th Armies on the western outskirts of the city.

The chosen destination for the relocation of KhPZ Zavod No. 183 was Nizhne Tagil in the Sverdlovsk Oblast, where it was to be merged with the already existing Uralvagonzavod (Ural Railcar factory) Plant, named after Felix Dzerzhinsky, the founder of the CHEKA. Also already on site was the Mariupol Ilyich works that had been evacuated the previous July. The new incarnation was renamed the Stalin Ural Tank Plant No. 183 and known by the abbreviation **UVZ UTZ No. 183**. In its new and its much enlarged form it would become the biggest producer of T-34s in the USSR. By 1945, UVZ had produced 28,952 of the 58,681 T-34s produced between the very end of 1941 and 1945. Retention of the Plant 183 designation signified that it was still the centre for T-34 GKB – the main T-34 Design Bureau. Indeed, within the plant the numbering of the many shops and departments were as they had been in Kharkov, with Morozov remaining

T-34 PRODUCTION FOR THE PERIOD 1940–45

	183 Kharkov	183 Nizhne Tagil	STZ Stalingrad	112 Gorki	174 Omsk	ChKZ Chelyabinsk	UZTM Sverdlovsk	Total
1940	115							115
1941	1,560	25	1,250	161				2,996
1942		5,684	2,520	2,718	417	1,055	267	12,661
1943		7,466		2,851	1,347	3,594	452	15,710
1944		8,421		3,619	2,163	445		14,648
1945		7,356		3,255	1,940			12,551
(1945 I)*		3,592		1,545	865			6,002
(1945 II)*		3,764		1,710	1,075			6,549
Total	1,675	28,952	3,770	12,604	5,867	5,094	719	58,681

1945 I* = production of T-34 through to end of May 1945.
1945 II* = production for remainder of 1945.
Note: *It is important to bear in mind that there are conflicting figures as to actual number of T-34s produced. Those above are certainly not the last word, but serve to give as a guide to the vast number of T-34s produced between 1940 and 1945.*

as head of the design team. Although Morozov and his design bureau returned to Kharkov in 1944, the UVZ retained the 183 number.

It took some three months for the factory to be fully relocated and for production to begin at the new site. In common with many other industries that had been evacuated, the new factory had to be created with limited materials, in many cases on virgin sites, with human muscle and bone employing hand-held tools to dig foundations, with people working day and night.

The scene of their work being lit by arc lamps or electric bulbs suspended from trees. . . . Even before the roof had been completed, the machine tools were already functioning. Even when the thermometer went down to forty degrees below zero [this being the worst winter Russia had suffered for over 50 years], people continued to work.

Notwithstanding the appalling conditions the workers at Plant 183 'were able to assemble their first twenty-five T-34 tanks, which were promptly sent to the front'. UVZ was also responsible for the development of the T-43 and the T-44.

In the hiatus caused by the removal of Plant 183 to Nizhne Tagil, T-34 production continued on just two sites, but at a very much reduced rate.

The **STZ** in Stalingrad produced just 1,250 in 1941 and the other site, at **Zavod No. 112** – the **Krasnoe Sormovo** (Gorki Shipyards – Red Sormovo) factory at Gorki, a mere 161 with the first of these being assembled by parts evacuated from the Kharkov plant in September. Zavod No. 112 had been ordered to begin production of the T-34 on 1 July 1941, and though it had been planned to produce a much larger number by the end of 1941, this proved impossible. The first few produced were delivered to the Moscow Front in November. Nonetheless, the period in which Plant 183 was non-productive marked the nadir in T-34 availability; the relatively small number emerging from STZ and Krasnoe Sormovo could in no way cover the demand for the machine from the Red Army.

In addition to the **UTZ**, the **STZ** and **Zavod No. 112** a further three factories were employed from mid-1942 in constructing the T-34. It was the threat of the loss of T-34 production from the Stalingrad plant that prompted this move.

The first of these was the **ChKZ** – the **Chelyabinsk Kirovsky Zavod** (Chelyabinsk Kirov Factory), **Zavod No. 100**. This factory was originally producing the KV-1 and KV-8 flame-throwing heavy tanks. The plant was ordered to tool-up for T-34 production in July

1942. This was completed in just 33 days with the first T-34 rolling off the line on 22 August. It continued with production of the T-34/76 and was the last plant to manufacture this variant until early 1944, by which time it had produced 5,076 examples of the model.

The second was **UZTM** – the **Uralsky Zavod Tyashelovo Mashinostroitelnya 'URALMASH'** or the Ural Works for Heavy Machine Building at Sverdlovsk. Ordered to tool-up for production of the T-34 in July 1942, the first 15 vehicles were completed in September. Some 267 were built in total for 1942 and a further 452 in the following year. Production of the T-34 was curtailed in favour of UZTM becoming the sole venue for the production of the SU self-propelled gun series – that is, the SU-122, the SU-85 and the SU-100.

The third plant to be converted to T-34 production in 1942 was the **Zavod No. 174 imeni 'K.E Voroshilov'** at Omsk in western Siberia. Production of the T-34/76 and T-34/85 amounted to 5,867 by 1945.

Each factory was at the nexus of a web of suppliers. As many of the sub-contractors for T-34 parts had not been evacuated, 1942 had seen the establishment of alternative suppliers and supplies to enable the T-34 to be built. As explained elsewhere in the book, this resulted perforce in a massive rationalisation of the number of parts and also of the materials used in the construction of each tank. By way of a simplistic example, the diagram below illustrates how the Krasnoe Sormovo plant in Gorki was supplied by feeder Zavods with elements of the T-34 which were then combined at the plant to create the finished tank.

Note: *The diagram naturally excludes a multitude of other concerns involved in the supply of the thousands of parts that went into building the tank.*

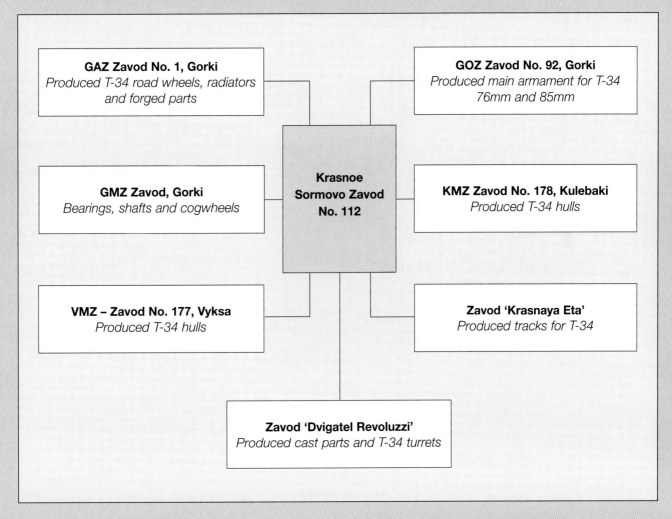

To oversee T-34 output at the UTZ 183 at Nizhne Tagil, which Stalin was of the opinion was too low (it had only begun production in late December after its evacuation from Kharkov), he decided to appoint a 'hatchet man' to gee them up. The picture that the Vozhd had received was that the factory – still in the process of being built – was 'in confusion'. He wanted someone in there to get the place sorted out and get production ramped up to the level he felt it should be at. To that end, in January 1942 he appointed a Col I.M. Zaltsman – clearly a man whom the Vozhd trusted – as director of the plant to smooth out the production process. What he said to Zaltsman is revealing about the way that even he was becoming aware of the effectiveness of the T-34 from reports from his front-line commanders. They sang its praises in consequence of its performance in the winter counter-offensive begun in front of Moscow on 5 December, which had thrown the Germans back at great cost to them in terms of men and equipment lost. The expansion of this offensive to embrace most of the Eastern Front in January 1942 for the first time permitted the T-34 to show how effective it was in dealing with the ferocious conditions wrought by the worst winter in Russia for 50 years and also onslaught of the German forces. Over the phone, he addressed Zaltsman as follows:

I know you like heavy tanks, but now please concentrate on the medium T-34 tanks. The Front commanders have called me on the telephone and asked to increase production of these agile vehicles that have performed so well in the deep snow and on the bad roads.

The manner in which the wide tracks of the T-34 conferred an ability to drive over deep snow without sinking was not matched by the narrower tracks of the Pz.Kpfw III and IV, which had the tendency to do just that. In consequence, they needed to stay on the roads – not that many in Russia were metalled – which it made it easier for the T-34s to attack them as they could travel across country and where and when necessary master the deep mud that many of these 'roads' were churned into by the passage of traffic. Having numbers of T-34s emerge out of the woods or the thick forests that flanked many roads in Russia and attack columns of retreating traffic did not serve to do anything positive for German morale.

But even as the Soviet winter counter-offensive was widened to embrace attacks all along the very long front, in a conference held in the Kremlin in January 1942 Stalin oversaw a decision that was to place a dead hand for the near term on any development of the T-34. In this matter he had been strongly influenced by these same very positive reports of the T-34s' performance he had received from various front commanders singing its praises. According to Malyshev, who had been appointed to the post of People's Commissar of the Tank Industry the previous year (and was to remain in that position through to 1956) and was in attendance:

Comrade Stalin emphasized several times that what was needed now was those weapons which had done well in battle; they were to be produced on a massive scale without making any modifications.

BELOW T-34/76 Model 1941's of the 116th Tank Brigade in service on the Western Front in May 1942. *(Gennady Petrov)*

RIGHT Tankers collect their new charges from the STZ works in Stalingrad. Another feature that marked out tanks produced at this plant was the manner in which the hull plates were connected by interlocked joints. *(Gennady Petrov)*

Malyshev also noted that Stalin said that this was a mistake the Germans had made. Indeed, Stalin would never have countenanced the loss of a whole month's output to accommodate the introduction of a new gun, as was to occur just two months later when in March all Panzer IV production was halted in Germany to accommodate the fitting of a new weapon. This was the model F2 that was being up-gunned to feature the 75mm L/42 weapon, designed to take on the T-34. Such a hiatus in production would have been anathema to Stalin. As the undoubted leader of the Soviet Union, as Supreme Commander of the Red Army and Minister of Defence, no one would have the temerity to gainsay him in this matter. In the circumstances, what he decided was understandable, but the continued 'freeze' in 1942 on any sort of significant improvement to the design was to be responsible for the erosion of the technological lead of the T-34 over its German opponents. What changes that were introduced were sanctioned because they increased production. Although the scale of the qualitative in German armour would not become fully apparent until 1943, the early technical advantages of the T-34, especially in firepower, would be fatally eroded as they deployed a new generation of more powerfully armed and armoured tanks and introduced significant improvements in firepower to older models. If the grail of Soviet industrial output had now become quantity at the expense of quality, it was the latter which was overwhelmingly the primary concern in German tank production.

Nonetheless, the degree to which Stalin's demands for the maximisation of output was realised, even in the face of the appalling conditions generated by the massive relocation of industries in the autumn of 1941 and winter of 1941/42, between 1 December 1941 and 1 May 1942, the tank factories engaged in production of the T-34 managed to deliver 2,049 new machines. Stalin now intended them to be put to good use in what transpired to be an operation that would be disastrous for the Red Army.

1942 – the year of 'Deep War'

The German Army was a hard taskmaster, but even in the face of the unprecedented losses experienced by the Red Army in 1941, lessons were being learned by the leadership of the Red Army as steps were taken to formulate a new doctrine for tank operations. The disbandment of the Mechanised Corps in July of 1941 now led to the creation of new Tank Corps to replace the smaller tank brigades that had been created in their stead. Initially four of these new armoured formations were created with each tank corps fielding 100 tanks of which 40 were T-34s, 40 T-60 light tanks and 20 heavy KV-1s. But within two months of their being established they had been enlarged, so that at the beginning of May 1942, each tank corps comprised three brigades with a total of 98 T-34s. Over the course of the year, 28 of these formations would be created. This was the first step in the establishment of a much larger formation – the Tank Army – the first two of which were raised on 25 May, with each fielding two of the new tank corps. In addition to the 3rd and the 5th, two more – the 1st and 4th – were formed in July but there was no

ABOVE A T-34/76 built by Plant No. 183 in early 1942. It has a welded turret.
(Gennady Petrov)

uniformity to their organisation with the average number of tanks on strength varying from 350 to 500 machines.

However, it was one thing to create the organisation with which they could seek to emulate their German opponents, but it would take a lot more to match their combat performance. This became all too apparent when the Red Army launched its disastrous Kharkov offensive in late May 1942. Although this is not the place to offer up a detailed account of this battle, it will be helpful to describe the salient points. (A more detailed coverage can be acquired by reading texts contained in the bibliography on the subject.) Stalin was convinced that when the Germans resumed their offensive, come the better weather, they would renew their bid to capture Moscow. Many of the German units deployed to Army Group Centre in April and early May were part of a deception plan named Operation Kremlin – designed to service and reinforce this Soviet perception. Their presence was but temporary, for they would in due course move south to be added to the German offensive forces already being concentrated in the Kharkov region preparatory to the launch of the Wehrmacht's summer offensive. Of the latter, the Soviets were unaware when the decision was taken by Stalin to order a 'spoiling attack' to capture Kharkov in early May. The expectation was that the Red Army units allocated to the operation would find only limited and weakened German forces, of which there would be just one panzer division whereas in actuality four were deployed in the vicinity of Kharkov.

Although the total number of German tanks was substantially lower than that assembled by the Red Army, the qualitative difference in favour of the former rested upon a far superior tactical organisation where combined arms *Kampfgruppen* (battlegroups), equipped and linked by sophisticated radio nets and with close air support from the Luftwaffe, permitted the Germans far greater flexibility in their operations. In total, the Red Army was fielding some 1,200 tanks with just 239 of those being T-34s, while the German total stood at 421 tanks. The Soviet offensive was launched on 12 May with the northern and southern forces intending to capture Kharkov by means of a pincer operation. After making initial headway, both forces had been brought to a halt by the Germans in less than a week and on 17 May they launched their counter-offensive. It was all over by the end of the month with catastrophic losses for the Russians. Total tank losses stood at 775 machines of which many were T-34s.

In the days after the Battle for Kharkov and prior to the launch of the German summer offensive in southern Russia, the General der Schnellen Truppen beim Oberkommando des Heeres saw fit to circulate a document which instructed panzer units on the Eastern Front as to how to combat the T-34. What is interesting is that in nearly a year since the invasion of Russia the Germans had still not managed to field in any sufficient quantity of new equipment that could deal with the Soviet medium tank. While it was certainly the case that the more powerful 75mm PaK 40 anti-tank gun was beginning to make an appearance, as was the 75mm L/43 in the Pz.Kpfw IV and StuG III (and from May onward the Pz.Kpfw IVG mounting the definitive 75 L/48 cannon), they were still too few in number to make any difference at this time. The substance of the document clearly implies that little had changed since 1941. The T-34 was still a much superior tank and was admitted to be such by the Germans. The document states:

The T-34 is faster, more manoeuvrable, has better cross-country mobility than our Pz.Kpfw. III and IV. Its armour is stronger. The penetrating ability of its 76.2mm cannon is superior to our 50mm Kw.K. and the 75mm Kw.K.40 as well as the 75mm Hohlgranate [hollow charge shells]. The favourable form of the sloping of all the armour plates aids in causing the shells to skid off.

Nor had the Russians been remiss in adapting their own tactics to exploit the superiority of the T-34. Bitter experience had taught them to take advantage of the evident superiority of the main armament, such that where possible T-34s would stand off and fire at the Mark IIIs and IVs at ranges between 1,200 and 1,800m. This was also aided by the greater speed of the Russian tank, which meant that in practice, if the T-34 was being operated by a moderately experienced crew, it could engage and break off contact at will – thereby choosing where and when to engage the Germans. This was also assisted by the T-34's superior cross-country performance because of its wider tracks. It was noted by the Germans that T-34s would often be dug in so that only the turret tops were showing, in order to facilitate an ambush. To do this, it was recognised by the Germans, that Russian tank crews would select a commanding height which permitted them to oversee the oncoming Germans, thus allowing them to employ their range advantage to the maximum. Other sites chosen for ambushes included the edges of woods where they could be difficult to spot and amid villages, where they would lie alongside the buildings and camouflage themselves.

It was therefore vital for the German panzers, if they were to survive their encounter with the T-34 in such situations, that they closed rapidly with the Russian tank. But to achieve this meant that the German tanks would have to close to a distance well within the firing envelope of the T-34. Whereas there was no point on the hull, suspension or turret that a T-34 could not penetrate either the Pz.Kpfw III and IV at or within the ranges indicated above, the German tanks not only had to advance to numbingly close ranges to the T-34, but their fire had to be directed at some very specific areas if they were to do it damage. The latter had only been identified through bitter and costly experience. These were formalised and issued to panzer crews and they had to be learned and remembered – a reasonable supposition given that the lives of the German tank crew would be at issue if not. In the original, they also covered how to deal with the KV-1. However, for our purposes this has been omitted.

Even while the German document quoted above continued to acknowledge the 'superiority' of the T-34 into 1942, it is worth taking note of a table published in a Soviet wartime study of those German weapons responsible for T-34 losses from June 1941 through to September 1942 expressed as a percentage of the total:

Pz.Kpfw IV with 75mm KwK L/24 gun versus T-34/76		
T-34	Penetrating effect	Hindering effect
Gr.38 HL	150–800m at the hull between 5th and 6th road wheel. 150–400m at the m/g ball-mount on glacis.	Jamming the turret. Damaging the gun. Damaging the track.
Pzgr	–	Same as above
Sprgr	–	At close ranges on the air outlet cowling at the rear.
Pz.Kpfw III with 50mm KwK L/42 versus the T-34/76		
Pzgr.40	Up to 200m at the hull between the 5th and 6th road wheel up to 100m as the turret side.	Jamming the turret. Damaging the track. Concentrated fire on the driver's visor possibly penetrates.
Pzgr	Up to 150m at the hull between the 5th and 6th road wheels after several hits.	Same as Pzgr.40
Sprgr	–	–
Pz.Kpfw III with 50mm KwK L/60 versus the T-34/76		
Pzgr.40	No experience since the shell casing frequently sticks in gun. Only fired at the KV-1.	
Pzgr	Up to 400m at the hull and turret side. At 300m from front at the driver's hatch after several hits.	Jamming the turret. Jamming the gun. Damaging the track.
Sprgr	–	At close ranges on the air outlet cowling on the rear.

Causes of all T-34 losses from June 1941 to September 1942 (expressed as % of losses)*								
Weapon calibre	20mm	37mm	Short 50mm	Long 50mm	75mm	88mm	105mm	Unknown
% lost	4.7	10	7.5	54.3	10.1	3.4	2.9	7.1

*It has to be borne in mind that it would not have been possible for the Russians to have known exactly what caused their losses in 1941. The exactitude of their figures is questionable, but nonetheless they are of interest.

The reference to the long 50mm is to the 50mm PaK 38 anti-tank gun, which as we have seen was in service from the beginning of the Russian Campaign, as well as the 50mm L/60 gun mounted on the Panzer IIIJ that started appearing on the Eastern Front from December 1941 onward and on the Ausf L which supplanted the previous model on the production lines in Germany from July 1942. Not part of either table is the impact of the effect of the new 75mm L/43 guns now making their appearance on the Eastern Front in small numbers on the Pz.Kpfw IV F2 and the StuG III F. Most of these were allocated to the formations earmarked for the German summer offensive. But in truth these would not make a really significant impact until 1943 when they finally became available in the numbers that made these larger calibres the norm for these tanks.

After the disaster of the Second Battle of Kharkov, which had seen many of the carefully husbanded T-34s frittered away in a very ill-conceived offensive, the depleted Russian tank formations had now to face the German summer offensive. Because of the scale of their *matériel* and personnel losses since June of the previous year, this was now concentrated in the sector of Army Group South. Given the codename of 'Case Blue', this operation was primarily directed at securing the oilfields of the Caucasus. The stated secondary objective was to bring the city of Stalingrad on the Volga under artillery fire, thereby neutralising it as an industrial centre and also denying the Russians the use of the river to transport oil supplies from the oilfields of the south. The Battle for Stalingrad as it subsequently occurred was not what was envisaged at the time Hitler issued his directive for 'Case Blue'.

For the Soviets, 1942 was the year of 'Deep War'. It marked the nadir of not just Soviet military fortunes but also of the quality of the machines leaving the production lines. Many T-34s were very poorly finished and so unreliable that they were frequently abandoned by their crews after very short journeys. One of the most graphic examples of how a vital resource shortage forced the tank plants to find some alternative to addressing the problem can be illustrated by the manner in which STZ responded to how the dire shortage of rubber impacted on their ability to fit rubber tyres to the road wheels of the T-34. Their solution was also then adopted by other plants. What they came up with in substitute was metal road wheels, which, while visually distinctive, were not at all appreciated by the tank crews who had to drive T-34s equipped with them. They were extremely noisy and could be heard by the Germans some distance away. They also generated harmonic vibrations in the machine, which served to loosen many parts. With the re-establishment of limited rubber supplies this problem was partly alleviated by employing

RIGHT A late-model STZ-produced T-34 distinguished by the undercut to the turret front and the hull sides which are devoid of any fixtures. The tank has been the victim of a fierce fire which has blackened much of the hull and road wheels. This photograph was taken in southern Russia, possibly in Rostov-on-Don or even Stalingrad. *(Gennady Petrov)*

rubber-tyred road wheels on the front and rear stations with the three in between still being fitted with the all-steel type. This arrangement could still be seen employed on T-34s manufactured in 1943.

Once again German photographs taken during the course of the summer offensive show many abandoned as well as destroyed T-34s. As in the previous year it was simply not the case that all had been knocked out by German fire. The issue of quality also profoundly impacted on that of the efficacy of the T-34's powerplant. As can be seen in Chapter 5, the motor was flawed by virtue of the very poor quality of the air cleaners which contributed to a very short engine life. Even in 1941 this was just 100 hours (those tested on the T-34 they received for evaluation by the Americans at the Aberdeen Proving Ground were described as being 'criminally poor' causing the engine to run for just over 70 hours before packing up). The average mileage to overhaul in 1942 was just 200km – meaning that few reached that figure and many broke down after having driven much shorter distances. Such was indeed the case with the fighting in southern Russia in the summer of 1942 where the huge expanses of the steppe caused tanks to throw up huge plumes of dust. Examples of this can be seen in both Russian and German film of the period. The panzers were not immune to the dust problem (it had been a major concern in the summer of 1941) but it impacted on the tanks of the Red Army far more with reports of T-34s in need of repair after just 10–15 hours of service. Many engines failed, their air cleaners taxed beyond capacity after just 30–50 hours of driving. Nor did the Red Army have the luxury of being able to recover and repair such machines (this applied to all of their tanks). Recovery vehicles were few and far between, so very few could be retrieved and repaired. In the summer of 1942, in southern Russia those that were abandoned for whatever reason became the booty of the Germans as they advanced in the course of their offensive.

Nor was it just the technical limitations of the design that impacted on the T-34 at this time. There was also the vital human dimension. A prodigious number of tanks were being produced – the total growing throughout the course of the year – even with all of the problems the facing the manufacturing industry with just over 12,000 examples of the T-34 being made in 1942. From a low of 454 tanks in January, this had risen to 1,568 in December. Such numbers also required that there be enough tank crews to man the machines, and quite simply, there were never enough. The short life of the tank was often matched by that of its crew, so the turnover was extremely high. This problem was in no way alleviated by what was clearly the very 'bone-headed' attitude wherein – and this was intrinsic to Red Army 'culture' – units committed to battle were all too often fought to destruction. Fighting to realise the objective, whatever the cost, was the way of the Red Army, with coercive 'encouragement' provided by the continual threat of court martial and execution or even summary execution at the hands of an immediate superior officer for failure. As the demand for crews increased to service the need for replacements in established units and also to man new ones being raised, this could only be realised by reducing training time – which was already remarkably short when compared with that given to German tank crews – with a corresponding decline in their effectiveness. Indeed, at its worst, the total length of instruction received by some crews could be measured in just a few hours. While a total of 34,664 tank crewmen were trained in 1942, this was a major shortfall in that the total number of all tanks produced in that year required no fewer than 82,000 crewmen. It took some while for the Red Army to come to an appreciation that tank crews could not and should not be treated in the same fashion as were infantrymen, who were always regarded as expendable. In this, the Red Army was no different in its attitude to the Army of the Tsars. It did, however, take until December 1942 for the order to go out that injured tank crewmen were not to be employed in units of other arms once recovered, but returned to service in the tank arm. An injured tank crewman was one who had survived, and was therefore 'experienced'.

The measure of how dire the situation was perceived by the Russians was that from the beginning of July – just a week after the beginning of the German offensive, through

THE RIGOURS OF T-34 TANK PRODUCTION IN WARTIME RUSSIA

There is no question that in 1942 the Red Army suffered its greatest shortage of weapons and equipment. Compounding the vast loss of the same in 1941 and the drop in output brought about by the huge dislocation in production (experienced owing to the relocation of industry to new sites to the east of the Urals), there came the serious problems generated by the loss of natural resources. This was because of the German occupation of much of European Russia and Ukraine.

The production of the high-grade steels needed for armour plate for tanks was severely impacted by the shortage of metals such as vanadium, chrome, nickel, molybdenum and manganese, and the loss of the factories of the Donbass in eastern Ukraine that had before the war produced most of the high-quality armour plate for tanks. Other sources of these ores had to be found – new mines for the extraction of manganese were opened in the Urals and in Kazakhstan and Sverdlovsk province. The large-scale smelting of ferromanganese at plants in Kushvisk, the Kuzbas and Magnitogorsk were matched by the development of molybdenum mines in the waterless steppe near Lake Balkhash in Central Asia. In consequence, there was a temporary lowering of the output of the blast and open furnaces.

But it was not just a question of reconstituting evacuated factories, but also of a 'fundamental reconversion of various production processes in the east', according to the author Alexander Werth. Compounding the wider matter of the supply of armour plate was the issue of the supply of subcomponents and the materials used to build them. At Uralvagon Plant 183, which had become the major T-34 manufacturer in 1942, this prompted a major review of what elements could be pared from the manufacture of the tank. In total, the process identified:

. . . the simplified production of 770 items and the total abolishment of 5,641 items of the tank. Another 206 items, previously shipped by subcontractors, were also recognised as unnecessary. In this light, the mechanical assembly time was reduced from 260 to 80 standard hours.

Among these was the removal of the oil radiator for the lubricating system, while the oil capacity was increased to 50 litres by way of compensation. Additionally, the rotary gear pump of the fuel system was replaced by a rotary lobular pump. The problematic nature of the supply of electronic equipment led to the deletion of a number of gauges and measuring systems, headlights, tail lights, electronic fan, horn and tank interphones not being fitted until the spring of 1942. Even the main armament was shorn of parts deemed to be superfluous to its primary function – that of firing a shell. The reduction from 861 parts to 614 while assisting output had an impact on the effectiveness of the gun in combat.

However, throughout 1942, there was a perception of a general decline in the quality of the T-34s being delivered. This was in consequence of the many changes designed to reduce production time and enhance output. In practice this was resulting in a situation where the tank was becoming virtually a 'one-use weapon' whose combat longevity could be numbered in days. Stalin's declaration to 'tankists' in mid-August of 1942 that abandoning their charges would be construed as cowardice and punished accordingly was a perverse response to a problem in part caused by his absolute insistence that the output of numbers of T-34s trumped any other consideration!

There is no historical counterpart to the immense programme of industrial relocation undertaken by the USSR in the latter half of 1941, the scale of which is still scarcely appreciated by the peoples of its Western allies. Neither is there any comparable knowledge or understanding of the role played by or of the appalling conditions endured by Soviet workers, many of them children, women and older people, as they strove to produce the weapons for the Red Army in the vast war economy in the Soviet Union. This was certainly true of many of those employed in the production of the T-34. They became the necessary substitute for the very serious shortage of manpower suffered by Soviet industry during this period. In 1940 there were 31.2 million workers involved in the national economy but with the onset of the massive call-up of manpower to swell the vast number of divisions being raised for service with the Red Army even before the German invasion, this had dropped to 27.3 million, plummeting thereafter in the months between June and December to just 19.8 million. Many of these also disappeared as countless workers were abandoned in the exodus to the east with the evacuation of the factories – numerous workers were taken, but not all could be. Yet even in the face of these appalling statistics and the massive dislocation of industry even then ongoing on 9 November, the GKO stipulated that in the following year, among the many other targets set down for weapons type production, no fewer than 22,000 to 25,000 heavy and medium tanks were to be produced.

Reconciling these realities meant that in those places where the vast new tank plants were to be constructed, and functioning by early in 1942, the workers to produce these could only be found among the young, women and the old as the military had first call on the manpower. For the civilian workers the incentive to work exceedingly long hours, day in, day out, and in many cases in appalling conditions, was that by doing so they could ward off starvation. A less emphasised consequence of the German occupation of Ukraine and other areas in European Russia was a catastrophic drop in the availability of foodstuffs. Here is not the place to explore this in detail, but quite simply for the old and the young of whatever sex it was a case of 'work or you starve'.

This can be illustrated by reference to the ChKZ plant at Chelyabinsk – better known as Tankograd – which had been ordered to begin building the T-34 in July alongside the heavy KV-1 tank already in production there. A newly built factory was created for this task and the workforce comprised 75% women and children under 16. This reality was mirrored throughout the Soviet war industries. It is typical that in the small number of photographs available showing T-34 production, most show children at work on the production lines. Alexander Sergeyevich Burtsev was commissioned as a junior officer in August 1944 and was sent to UVZ Plant 183 for tank training and to be allocated a crew. They found themselves working in the factory itself and although not very old himself – just 19 – he expressed surprise at the age of some of the workers.

He recalled:

. . . Here is your tank. We put the rollers on alongside the factory workers, and helped as much as we could. High class specialists worked in the tank assembly, and there were some really young guys – drivers 13–14 years old were there. Imagine a huge workshop, and tank assembly lines to left and right of you, and a tank rushes in between at a speed of thirty kilometres per hour with a kid like that at the levers! You just couldn't see him! The tank was about three metres wide and the gate a few centimetres wider, and the tank dashes through the gate at that speed, flies onto a platform and stops dead!

In the same way that the Red Army employed women on an unprecedented scale compared to the Germans, the Americans and British, and thus made a major contribution to winning the war in the East, it is also very clear that the role of young people and women working in the factories assembling T-34s en masse made a major contribution to helping secure the victory over Germany.

RIGHT A knocked-out and abandoned STZ-constructed T-34 in Stalingrad. *(Gennady Petrov)*

BELOW A superb image for illustrating two of the new *Gaika* or hex-nut turrets developed by Plant 183 in mid-1942 and adopted by all other plants shortly thereafter. Designed for ease of manufacture it still housed just two men. The T-34 Model 1943 in the foreground was produced by the 183 factory and employs a mixture of road wheel types. Road wheel station Nos 1 and 5 have rubber tyres, whereas the others are of the all-steel type. Note also the box-type fuel tanks on the hull rear, although these were also a characteristic of those T-34s produced by Zavods No. 112 and No. 174 through to the spring of 1943. The T-34 in the background employs a drop-forged turret originally produced by URALMASH although then employed on the hulls produced by Plants 112 and 183. It is for that reason more difficult to identify the builder with exactitude. Note that it also mounts the round, cylindrical fuel tanks which were also produced first on URALMASH-produced machines from late 1942. The drop-forged turret was the largest of the three types of *Gaika* turrets. *(Gennady Petrov)*

to September – total Red Army tank losses across the whole of the Eastern Front were some 8,000 machines. It is often overlooked that combat was also continuing on other sectors of this massive front and this number did not obviously just comprise T-34s, but also included a much smaller number of KV-1 heavy tanks and large numbers of various types of light tanks. These numbers betoken an ever-present reality of tank warfare in the East – that Soviet losses relative to those of the Germans was always far, far higher. However, it must be borne in mind that while German propaganda might imply that every Soviet tank photographed or filmed had met

its demise at the hands of a victorious panzer crew, the truth, as we have seen, was more prosaic, in that very many of those T-34s found abandoned were victims of inherent technical problems and not German firepower.

The retreat of the Red Army in the face of what was at the time seen as the inexorable German advance in southern Russia and the near panic engendered thereby prompted the Stavka (Soviet High Command) to issue Order No. 228 – the infamous *Ni Shagu Nazad* or 'No Step Back' pronouncement of 30 July. This was reinforced by their communication to tank troops in mid-August following the defeat inflicted by the German 6th Army on the Soviet forces defending the approaches to Stalingrad at Kalach on the River Don. In this encirclement battle, the Red Army lost some 1,000 tanks belonging to seven armoured brigades and the two motorised brigades of the recently half-formed 1st and 4th Tank Armies. Its coercive tenor was all too familiar:

> . . . *Our armoured forces and their units frequently suffer greater losses through mechanical breakdowns than they do in battle. For example, at Stalingrad front in six days twelve of our tank brigades lost 326 out of their 400 tanks. Of those 260 owed to mechanical problems. Many of these tanks were abandoned on the battlefield. Similar instances can be observed on other fronts. Since such a high incidence of mechanical defects is implausible, the Supreme Headquarters sees it as covert sabotage and wrecking by certain elements in the tank crews who try to exploit small mechanical troubles to avoid battle.*

Nor could crews who were deemed to have abandoned their charges in the way that Stalin described expect to receive anything but punitive treatment from the military authorities:

> *Henceforth, every tank leaving the battlefield for alleged mechanical reasons was to be gone over by technicians and if sabotage was suspected, the crews were to be put into tank punishment companies or 'degraded to the infantry' and put into infantry punishment battalions. Inasmuch as some measure was required to determine whether a tank that had been abandoned had been so for a justifiable reason, the criterion became that of whether it was seen to be on fire, or at least smoking, after being hit by the enemy.*

Having captured Rostov-on-Don, German armoured units pushed south into the Caucasus with the intention of securing the southern oilfields. After the defeat of the Soviet forces defending the approaches to Stalingrad, the 6th Army began its advance towards the city and reached the River Volga on 23 August. During this time the STZ was producing T-34s as rapidly as possible with the last machines reportedly leaving the factory and joining combat in an un-primed state in September. Such was the scale of the heavy fighting in the city that the STZ was only finally captured by the Germans on 15 October. Given the assumed threat to T-34 production caused by the German advance on Stalingrad, the decision had been taken in advance to start up production on three other sites to the east of the Urals as the loss of STZ would otherwise have been very keenly felt – the plant having

BELOW A T-34 built by Plant No. 183 and in full winter whitewash oversees a desolate scene of snow on the eastern Ukrainian steppe, post-Stalingrad. *(RGAKFD in Krasnogorsk via Stavka)*

LEFT Another STZ-produced 'locomotive-wheeled' T-34. Given the proximity of the STZ works to the front in southern Russia in 1942, many of this model of T-34 were encountered by the Germans. Noteworthy again is the offcut under the front turret. STZ was captured and destroyed by the Germans on 15 October 1942. *(Author)*

been responsible for some 42% of total T-34 production up to the time it was destroyed.

Only one of these, Zavod No. 174 at Omsk, which began production of the T-34 in mid-1942, would continue with it until the end of the conflict, by which time it was to produce 5,867 T-34/76s and T-34/85s. The other two sites, namely the ChKZ or the Chelyabinsk Tractor Factory ('Tankograd'), produced T-34s from August 1942 through to March 1944 with the other being the UTZM, better known as 'URALMASH'. It was, however, to be involved in T-34 manufacture for only a limited period, with full production beginning in the midsummer of 1942 and ending in the autumn of 1943. The decision to terminate T-34 production at this site was taken to permit URALMASH to concentrate on the manufacture of the SU series of assault guns and tank destroyers for the Red Army. The beginning of 1942 witnessed the nadir in T-34 production, with just 454 tanks produced and mainly from the STZ with a small number from Zavod No. 112. However, by year's end, and even with the STZ destroyed, the total produced by all six other factories engaged in T-34 production was a staggering 12,661 new-build machines. Nonetheless, total T-34 losses in that same year amounted to some 6,600 machines, which was considerably in excess of

LEFT Although the myth was often offered up that the T-34 was less likely to catch fire because it was fuelled by diesel, the truth, as can be seen from this image, was otherwise. The proximity of the fuel tanks along the inside hull sides carried a very definite fire risk if the hull sides were penetrated. Indeed, exposing the hull of one's T-34 tank to the enemy became an offence that could lead to execution for negligence. *(Author)*

all tanks lost by the Germans in their disastrous 1942 campaign.

During this time, an upgrade was introduced on the T-34, which, while not detracting from Stalin's prohibition against any change that impacted on productivity, saw a variation in the T-34/76's appearance. Although there were a number of internal changes, the most visually apparent was the introduction of a new turret – its appearance first being noted by the Germans in July 1942. The new hexagonal shape had been designed at Plant 183 by M.A. Nabutovsky for the specific purpose of making production of the turret easier and speedier. But despite the change to its outward appearance, it did not address the fundamental problem – it still housed just two men, even if they did have slightly more space. Although the design was taken up by all of the factories producing the T-34 – the changes having been sanctioned for manufacture by the GKO on 1 July 1942 – as with the earlier turrets, not all were manufactured in the same fashion. There were three main types. Two were cast and one was stamped. However, the methods employed to produce them reflected the particular limitations, equipment and expertise of the plants concerned. Examples of the former, no matter how they were produced, were differentiated by being designated either hard-edge or soft-edge. This reflected the number of sections employed by the plant in producing the turret, with as many as ten parts being welded to form the finished product in one case, while elsewhere, just two parts were used, namely the turret body and the turret ring. In all cases bar one, the top of the turret was welded to the body. The most distinctive and largest of these new turrets was produced by a process unique

to UTZM and made its appearance in October 1942. A 5,000-ton press had been purchased from Germany before the war and installed at the factory. This was now used to stamp out each turret from a single sheet of 45mm-thick armour plate which was then welded to the turret ring. In addition to producing the turret for ChKZ, URALMASH also supplied these stamped turrets to Plant Nos 112 and 183. By March 1944, URALMASH had produced and supplied 2,670 turrets. T-34 tankers already had a slang name for their charges, referring to it as the *Tridtsatchverka* to which they now added the name *Gaika* meaning 'hex-nut', after the shape of the new turret design.

The encirclement of the German 6th Army in Stalingrad in November 1942 and

ABOVE At the tail-end of the winter of 1943 the Germans inflicted a defeat on the Russians in the Third Battle of Kharkov. It was their last victory in the East. This T-34 Model 1943 has been knocked out on a street in the city in the very intensive fighting that took place when the Germans recaptured it. *(Author)*

RIGHT The introduction by the Germans of their Tiger I heavy tank mounting an 88mm gun in the late winter of 1942 was one of the primary factors in forcing the hand of the Red Army to improve the firepower of the T-34/76. In tests at Kubinka using a captured Tiger it was demonstrated that a 76mm F-34 gun could not penetrate the frontal armour of the heavy tank. At Kursk the Germans deployed over 100 Tigers and they did great damage to Russian tanks. *(Author)*

its subsequent destruction in February 1943 really did signal that the Germans could not defeat the USSR, although it took the Battle of Kursk in July 1943 to make that definitive. The Soviet counter-offensives during the winter of 1942/43 pushed the Germans back almost to the positions at which they had started their summer offensive in June. The Red Army received a bloody nose from a revived German tank force in the Third Battle of Kharkov, but this marked the last victory of German arms in the East. But in the late winter Soviet tank forces in the south of Russia had encountered a new German heavy tank whose armour and firepower were to threaten profoundly the T-34/76.

The first Tiger I heavy tanks saw service in Russia towards the tail-end of 1942. They were a rarely encountered beast for most Soviet T-34 tankers, although when they were it rapidly became apparent that this new German machine was a profound threat to the Russian medium tank. Just how evident this was can be gauged from the following account of the encounter of two Tiger Is belonging to the heavy company of the elite German Army division *Grossdeutschland* with a unit of T-34s in February 1943. The account comes from the memoirs of Erhard Raus and describes events that took place during the course of the Third Battle of Kharkov. It is worthy of mention simply because it shows graphically how much the reliance on the quantity production of the T-34, at the expense of any sort of real qualitative improvement in the design, had now rendered it extremely vulnerable to the first of a new generation of German panzers:

> . . . *two Tigers acting as a panzer spearhead destroyed an entire pack of T-34s. Normally the Russian tanks would stand in ambush at the hitherto safe distance of 1,200 metres and wait for the German tanks to expose themselves upon exiting a village. They would then take the tanks under fire while our Pz.Kpfw IVs were out-ranged. Until now, this tactic had been fool-proof. This time, however, the Russians miscalculated. Instead of leaving the village, our Tigers took up well-camouflaged positions and made full use of the longer range of their 88mm guns. Within a short time they had knocked out sixteen T-34s that were sitting in open ground, and when the others turned about, the Tigers*

BELOW T-34s were available in very large numbers at Kursk where they suffered extensively at the hands of German armour, most of which were employing guns that could defeat the frontal armour of the Soviet medium tank at battle ranges. *(Gennady Petrov)*

pursued the fleeing Russians and destroyed eighteen more tanks. Our 88mm armour-piercing shells had such a terrific impact that they ripped off the turrets of many T-34s and hurled them several yards.

In a matter of months, the Tiger I would be joined by the new Panther medium tank whose 75mm L/70 weapon could also execute such damage to the T-34, but unlike the heavy tank, was expected to be produced in far larger numbers for the panzer divisions. It was after all intended to become the standard medium tank of the Panzerwaffe, an expectation never realised. The year 1943 would then witness the demise of whatever qualitative superiority that still resided in the T-34/76 design, thus providing the necessary incentive to the Soviet tank industry to upgrade the design and enable it to survive on an increasingly dangerous battlefield. It would, however, take the seminal experience of the Battle of Kursk in July 1943, to drive home that need to Stalin and the Soviet tank industry. Quantity of production would no longer be enough to suffice.

Although it did not affect production, an attempt to rectify the limitations of the firepower of the T-34/76 in the face of these new and improving older German designs was the introduction of a new 'special' APDS (armour-piercing discarding sabot) shell to supplement the standard BR-350A anti-armour shell that had been in use since 1941. Allocated the designation BR-350P, this new shell had a higher performance than the 350A in that it could penetrate 92mm of armour at 500m, compared to 69mm at the same distance. Even this 'special' shell, however, would not permit the T-34/76 to penetrate the 100mm frontal armour of the Tiger I at the same distance. The only way that the Tiger and Panther could be rendered vulnerable was by hitting them in their flanks where their armour was thinner – in the case of the Tiger 80mm, and the Panther at 45mm. However, the T-34 had to close the distance before it became a victim to the latter's much longer ranging main gun. Even before Kursk, it was apparent that the T-34 needed a bigger gun but it would take the heavy losses of tanks at Kursk to finally wake up both Stalin and the High command of the Red Army to that need.

T-34 AND PANTHER – PRODUCTS OF DIFFERENT INDUSTRIAL AND WAR-FIGHTING CULTURES

In general, although praise was expressed by the Germans for the T-34, there was much about it that was thought to be poorly designed and poorly produced. Certainly, they were more crudely finished than the products emerging not just from German tank factories but also from those of Great Britain and the USA. Descriptive comments such as 'rough job', 'sloppy finish' and 'crude and poor welding' all figure in the reports both countries issued on the T-34 they received for evaluation. These sentiments would have been totally agreed with by the Germans. The quality of finish was of a level that would never have emerged from a German tank factory such as Krupp-Gruson, MNH or MAN. And while the Germans impressed a number of T-34s into service as *Beute-Panzers* on the Eastern Front (and deployed a few elsewhere) they would never have been acceptable as home-produced vehicles for service with the German Army. At issue between these two warring powers was a fundamentally different approach to war-fighting and industrial production.

German industrial culture, with its perennial concern for the quality of its product, was reflected in the weaponry produced for the Wehrmacht. This was something laid down by the armed forces themselves. In a lecture delivered at the War Academy in 1936 the speaker stated that (for the German Army): 'Armament for us is a question of quality in every sense, but particularly in the technical sense . . . an armament that overemphasises quality and speed of production is only to be achieved at the expense of material and industrial quality.'

The seemingly modern notion of *vorsprung durch technik* thus has quite a long provenance! According to Albert Speer, as German Armaments Minister, this obsession with the quality of product and all things to do with it prompted the German Army to remain hostile to his suggestions, even after the fortunes of war had turned against Germany, that they should simplify the design of their weapons so as to accommodate mass production. A shared perception of US, British and Russian evaluation teams when examining captured German tanks was that many of them were over-engineered, either in whole or in part, and too finely finished given the exigencies of the type of conflict Germany was facing from 1941 onwards. Even small items like weld beads on the hull or those on turret bins were ground down to a very high tolerance, while more complex items of equipment such as engines and transmissions were finely honed. ⇨

Quite simply the Russian perception, as is reflected in the close-up images of the very rough finish of the T-34/85 illustrated in the 'walk-around' on pages 125–129, was that unless this happened to intrude on the effective working of the tank and its crew – then leave it be. To do otherwise would only waste time and resources. It did not matter that they looked crudely finished – war is not a competition in aesthetics! This comes back to a point made earlier and is reflective of all Soviet tank production, even in the modern period, that 'the best is the enemy of the good', where the good is massive numbers. In that for many, certainly in the last two years of the conflict, the T-34/85 and the German Panther symbolised the clash of the mediums, it might be fruitful to compare and contrast aspects of the two designs.

The Soviet view was that the Panther was not only too heavy for a medium tank, it was also unnecessarily complex. By the Soviet weight classification for tanks, the 45.5 tons of the Panther Ausf A placed it in the heavy tank category. Even the IS-2, which was the Red Army's primary heavy tank for the final year of the war, weighed in at 44 tons. That is over a ton less than the medium Panther. In actuality, the overcomplexity of the Panther manifested itself in its unreliability, which was prevalent in all of its production iterations. That meant that at any given time, a significant number of those in service on the Eastern Front were out of action awaiting repair. A tank under repair or awaiting repair is a tank that cannot fight, no matter how 'effective' its specification on paper. The Germans, for example, never managed to solve the ongoing problems with the Panther's transmission before war's end.

In addition to those factors already mentioned, Russian criticism of the Panther also took into account the contrast between the cost of each machine and the man hours taken to construct either type. By 1943, when production of the T-34 was at its height with 15,710 being made, the cost to the Soviet state of building each one had been reduced to 135,000 roubles – the equivalent of $25,470, while it took 3,000 man hours to manufacture. Compare that to the Panther, where each machine cost the Reich about 129,000 Reich marks approximating to $51,600 and 55,000 man hours to produce!

In theory then and in terms of man hours, in the time it took German workers to produce one Panther, Soviet workers could produce many more T-34s.

For the Soviet Union, the primary concern in tank production was never quality; it was always about quantity. Quality only became an issue as it did in 1943 when it was obvious, even to Stalin, that superior numbers could not compensate in the face of new and superior German tanks. Only then was something done to enhance the survivability of the T-34. The Soviet way of war was always about the power of numbers, whether it was in manpower, or, given our concern here, tanks. The Russian Army had historically always been a 'meat grinder', now rendered even more bloody and effective by modern military machines. The military writer P.H. Vigor noted that Soviet military doctrine had always been based upon this premise:

> *In Soviet (military) thinking the concept of economy has little place. Whereas to an Englishman [or German] the taking of a sledgehammer to crack a nut is the wrong decision and a sign of mental immaturity, to a Russian the opposite is the case. In Russian eyes, the cracking of nuts is clearly what sledgehammers are for.*

Whether or not it was Lenin or Stalin or someone else who was responsible for the expression 'quantity has a quality all of its own', it is apposite to the case being put here. Herein lay the very essence of the Soviet way of war. From the outset, the T-34 was a tank that was always intended to be built in the thousands, and, as we have seen, from the beginning of the Russian tank industry in the early 1930s it was a given that tanks were expected to be manufactured in such numbers. The corollary to this was that in a war of masses – be it in manpower or tanks – losses too would be huge. Indeed, the life of a T-34, be it a T-34/76 or the later T-34/85, was always assumed to be short, in many cases numbered to just a few days and if lucky, possibly weeks. As to the question which was the better tank – the T-34 or the Panther? – the only response that a Russian would give would be that of a question: Who won the war?

The Battle of Kursk–Orel

The high-water mark of the T-34/76's combat career came when it helped defeat the last German summer offensive – codenamed Operation Citadel – directed at excising the Kursk salient in eastern Ukraine, in July 1943. It was also the battle in which for many Soviet citizens the Red Army symbolically came of age and is represented by the triumph of the T-34/76 in bringing about the defeat of the enemy. But paradoxically it also revealed beyond doubt that unless some redress was attempted, the main tank of the Red Army had now become extremely vulnerable and in its 76mm armed form, was verging on the obsolescent.

By that date, the bottlenecks and many of the component shortages that had severely impacted on the build quality and output of the T-34 in 1942, had been mostly overcome. The T-34 was a more mature tank – though still far from perfect – and the crews who engaged the Germans in the battle were some of the very best trained of the conflict. The T-34 was also emerging as a 'universal tank' in the Red Army, having eclipsed the importance of the KV-1, with the numbers available for operations at Kursk and in the planned counter-offensives to be launched once it was certain that the German offensive drive had been defeated, numbering in the thousands. It was indeed, the mainstay of the Red Army.

What was even more remarkable was that for this operation the Germans were fielding new tanks such as the Panther – which was to make its combat debut in the battle – and over 100 Tiger Is, both of which were superior in armour and firepower to the T-34/76. Also being used for the first and only time in any number, was the *Panzerjäger* 'Ferdinand'. This was a heavily armoured and armed mobile fortress mounting the PaK 43 88mm anti-tank gun that could defeat any Red Army tank at long range. In addition, many of the infantry formations had lightly armoured *Panzerjäger* attached, which were mounting the 75mm PaK 40 anti-tank gun. Apart from the long-barrelled 50mm mounted by the Panzer III, which although serving in the largest numbers of any German tank in the battle (this was nonetheless to be its swansong as a main battle tank) and the

ABOVE Although the Pz.Kpfw III was already out of production as a battle tank by July 1943, its chassis continued to be used to form the highly effective StuG III assault gun. Mounting a 75mm L/48, it was a very successful tank destroyer even though not designed explicitly for that role. It was encountered more frequently in 1943–44 as it was both cheaper and easier to produce than a tank. *(Author)*

few surviving Mark IIIs and Mark IVs carrying the short-barrelled 50mm and 75mm gun, every other Panzer IV, StuG III and *Panzerjäger* mounted the long-barrelled 75mm KwK40 L/48 cannon which could now defeat the T-34 at greater ranges than it could defeat these German machines. Thus for the first time since the onset of the war, the T-34 was operating at a huge qualitative disadvantage to most of the German AFVs encountered at Kursk.

However, notwithstanding the importance of the large numbers of T-34s and other tanks deployed within the salient, the basis of the Soviet plan to defeat the Germans was the creation of a highly sophisticated and massive defensive system with fixed emplacements, screened by huge minefields and supported by an immense amount of artillery. The intention was to snare and impale the German armour on and within these defences and shoot them with the thousands of artillery pieces deployed throughout the defensive system. The hundreds of T-34s and supporting T-70 light tanks and few KV-1 heavy tanks would be employed to help erode the large numbers of panzers and assault guns deployed as the cutting edge of the German offensive.

ABOVE The losses of T-34s at Kursk and in the follow-up Red Army counter-offensives finally drove home to the Red Army and Stalin that the T-34 had to be updated – and that meant a new gun. It was only after the defeat of the last German summer offensive and the price that had to be paid to achieve it, that a real urgency was lent to the task. *(Central Museum of the Armed Forces, Moscow, via Stavka)*

BELOW Kursk marked the combat debut of the new German Panther medium tank, the machine that had been designed to take the place of the Pz.Kpfw III made obsolete overnight by the T-34 in 1941. Its 75mm L/71 flat-trajectory main gun could 'kill' a T-34 at long range and although deemed a failure in the battle because of mechanical faults – it having been committed to battle too early – it nonetheless was responsible for knocking out more Soviet tanks than any other German type. Over 6,000 would be built by 1945 and would prove to be the principal nemesis of the T-34 through to the end of the war. *(Author)*

To ensure the success of Operation Citadel in the face of the extensive and deep Soviet defences constructed on either neck of the salient and also defeat the large number of tanks – primarily T-34s deployed by the Russians within it – Hitler ordered that there be a massive assemblage of German armoured might. To this end, the Germans garnered no fewer than 2,450 tanks and assault guns – a high proportion of the total number of AFVs available for service on the whole of the Eastern Front. As initially formulated, the offensive had been due to be launched as soon as the ground had dried out enough to permit the operation of armour on a large scale. It had also been presumed by the Germans that they would be attacking Soviet forces that were themselves assembling to launch their own offensive, but it soon became clear that this was not the case. Stalin had been weaned off this option and persuaded to embrace the idea of a strategic defensive operation by transforming the Kursk salient into a massive defensive bastion through the laying of vast minefields, the construction of large numbers of defensive positions, the assemblage of huge concentrations of artillery and the build-up of very large numbers of tanks.

Hitler, however, did not surrender the idea of an offensive, now seeing in the infusion of large numbers of more tanks and assault guns – especially of Tigers, the Ferdinands and particularly the Panther, that was to make its combat debut at Kursk – the solution to breaking the massive Soviet defensive system. The Panther was still experiencing many teething problems and it was to accommodate the time needed to deal with them that the launch date for Citadel was repeatedly postponed. The hiatus caused by German hesitancy – they were unsure how to react to the overwhelming evidence that the enemy was preparing to fight a defensive battle – now played more into the hands of the Russians than to the Germans. With factories in the east producing nearly 2,000 T-34s per month and the virtual cessation of major combat operations across the rest of the Eastern Front, this permitted the Red Army to build up a huge number of T-34s that were to be employed first to defeat the German assault within the salient and thereafter participate in a series of counter-offensives after it.

Operation Citadel – 5–17 July 1943

ABOVE T-34 1943s of a Guards Tank Unit serving in the Kursk salient, commanded by Col. V.V. Sytnik, July 1943. *(RGAKFD Krasnogorsk via Stavka)*

When the offensive finally began in the early hours of 5 July 1943, in the north of the Kursk salient the German forces were directed against those of the Central Front, while those of Army Group South, which was the stronger of the two German formations, was facing the Voronezh Front. Unlike in the north where the German commander released his armoured divisions on an individual basis, those in the south were launched at the Soviet defences en masse and from the outset of the operation. Numbers and deployment of Red Army tanks within the salient are given in the table above right.

Three Panzer Korps, which collectively amounted to just over 1,000 tanks including 194 Panthers and 57 Tiger Is, attacked the Soviet defences and while the two on either flank experienced delays, the SS Panzer Korps in the centre broke through the first line of defences within a matter of hours. This prompted the move forward of 1st Tank Army. By the day's end, the order had also gone out to 5th Guards Tank Army, laying in reserve some 350km to the east, where it was waiting to be used in post-Kursk operations, to move west for operations in the salient, as it was feared that the SS Panzer Korps might break through the remaining defences before Kursk. This saw a further 850 tanks – of which 501 were T-34s – earmarked for combat in the salient.

In the north, the German rate of advance from day one of the operation was slow, as

Front	Tanks in service at Kursk (excludes Lend-Lease machines)			Total
	T-34	KV-1/1S	T-60 and T-70	
Central	924	70	587	1,581
Voronezh	1,109	105	463	1,677
Total	2,033	175	1,050	3,258

BELOW The Soviet tank forces at Kursk were composed of a whole variety of T-34 models. These Type 1942 variants (the type evaluated by the US and British teams in 1943) were still serving in some numbers. *(Gennady Petrov)*

ABOVE A full two years after the T-34 had revealed the Panzer III to be obsolete as a medium tank it was nonetheless the most numerous German panzer to serve in the battle. This Mark III belongs to the 11th Panzer Division that served in the south of the Kursk salient. *(Gennady Petrov)*

the infantry, supported by the growing number of panzer divisions fed into the operation on a piecemeal basis, ground their way forward through the deep defences under massive enemy artillery fire and assailed by the tanks of 2nd Tank Army. It was the commitment of this formation on 6 July that initiated a tank clash every bit as large as that of the more famous one at Prokhorovka several days later. Ivan Sagun, the commander of a T-34, recalled many years later his encounter with a Tiger in this battle:

> . . . he fired at me from literally one kilometre away. His first shot blew a hole in the side of my tank. With his second, he hit my axle. At a range of half a kilometre, I fired at him with a special calibre shell, but it bounced off him like a candle. I mean, it did not penetrate his armour. At literally 300 metres, I fired my second shell. Same result. Then he started looking for me, turning his turret to see where I was. . . .

At that point, determining that discretion was the better part of valour, Sagun ordered his driver to reverse the tank and seek the cover of trees. He lived to fight another day. This short account illustrates the real problem that the 76mm-armed T-34 now had against these much more powerfully armed and armoured German beasts. By 10 July, the German assault in the north had been effectively stalled. Even the Tigers and 90 Ferdinands had not provided the Germans with the strength needed to break the massive Soviet defences.

Between 6 and 11 July, the southern thrust also ground its way forward. The 48th Panzer Korps on the western flank found its ability to advance continually hampered by the almost constant attacks from Soviet armoured units coming in from the west. It was here that the Panthers were deployed, but their number dwindled considerably over the first week. General Guderian, the Inspector General of German Panzer troops, was present on 10 July and commented on the sheer number of T-34s in action, likening them to 'rats streaming across the landscape'. On the eastern flank, 3rd Panzer Korps had finally made the breakthrough it should have achieved days before, and was now moving north to assist the SS Panzer Grenadier Divisions as they began their turn to the north-east in the direction of Prokhorovka.

Notwithstanding the number of breakdowns of T-34s en route to the battlefield, it was 5th Guards Tank Army that reached the town of Prokhorovka first on 11 July. Overnight it made preparations for its encounter with the SS Panzer Korps the following day. By this date, the 1st

ABOVE These two T-34/76 Model 1942s equipped with the Gaika turret have been fitted with PT-3 mine-rollers that first saw service with Red Army in the Battle of Kursk. *(Gennady Petrov)*

Tank Army had been reduced to a shell of its former self, having lost most of its tanks over the previous six days of heavy fighting. Fierce fighting also took place to the south of Prokhorovka as Soviet tank units managed to block the advance by 3rd Panzer Korps of Army Detachment Kempf, thus denying its tank strength as reinforcement for the SS Panzer Korps.

In the early morning of 12 July, 5th Guards Tank Army began its advance hoping to catch the Germans before they had assembled for their attack only to run headlong into the panzers of the 1st SS Panzer Grenadier Division that had begun their advance on Prokhorovka, more or less at the same time. Knowing that they were at a range disadvantage to the

LEFT The Pz.Kpfw IV had emerged as the main battle tank of the Panzerwaffe after the appearance of the T-34 in 1941 had rendered the Pz.Kpfw III obsolete. Never intended for this role, its ability to take a bigger gun saw it improved many times and remained in production and service with the German Army through to 1945. Its 75mm L/48 gun could penetrate the frontal armour of the T-34. *(Author)*

German tank guns, the Russians intended to close the distance as quickly as possible in order to get in among the panzers and hit them in the flanks. It was unfortunate for the Soviets, however, that the Germans caught them when still at a distance where they could fire and defeat the T-34s and the accompanying light tanks. In a ferocious hour-long engagement across the undulating terrain between the River Psel to the north, and the Kursk–Orel railway line to the south, the Russians lost upwards of 60% of their armour as total write-offs. Although it is this confrontation that has become synonymous with the Battle at Kursk, and while German losses were far less than those of the Russians, the outcome must nonetheless be construed as a Soviet victory. By day's end, the Germans had not succeeded in capturing Prokhorovka, nor did they do so in the remaining days before the offensive was closed down. Indeed, within days, Citadel was ended at Hitler's behest as on 17 July the Russians began new offensives to the north and south of the salient, forcing the German leader to draw upon the panzer divisions within it to contest them. Although Soviet tank losses were extremely high, this was the price that Stalin and his generals knew would have to be paid to see the German offensive worn down and defeated. The great prize of such sacrifice, however, was the final and irrevocable shift of the strategic initiative towards the Red Army.

The need for a T-34 with a bigger gun

Stalin's veto against modernising the T-34, if that resulted in the reduction of the numbers leaving the production lines, continued even beyond the point that it had become apparent that the Germans were deploying new and up-gunned tanks, assault guns and anti-tank guns that were serving to reduce the effectiveness of the Russian tank in combat. But it took the experience of the Battle of Kursk–Orel to finally motivate Red Army authorities to begin a programme to up-gun the T-34.

The Kursk–Orel defensive operation and the follow-up Kutuzov and Polkovodets Rumyantsev counter-offensives had proven to be a salutary experience for the Red Army, with the numbers of tanks –primarily T-34s – destroyed being very high. The Russian historian G.F. Krivosheev calculated that in the period between 4 July 1943, when the German offensive began, and 23 August, the date on which the Red Army officially deemed their 'Kursk' operation to have concluded, they had 6,064 tanks and assault guns (SU-76s and SU-122s) 'written off'. This averaged out to a loss rate of 121 per day. Nonetheless, in July and August some 4,000 tanks – the bulk of them T-34s – had left the production lines, and these, in addition to the immense reserve of T-34s built up by the Red Army in the period prior to the launch of Citadel, enabled the Soviets to maintain the high tempo and wide-ranging nature of their counter-offensive operations. Thus when the counter-offensive to liberate Belgorod and Kharkov was launched on 7 August, 1st Tank Army was fielding 542 and Rotmistrov's 5th Guards Tank Army 503 tanks, with the majority of these, in either case, being T-34/76s. A total of 2,832 tanks and assault guns were available to the three participating fronts on the opening day of Rumyantsev, giving the Soviets a 5:1 advantage in armoured vehicles over the German forces. This was just three weeks after the Germans had terminated all offensive action in the Kursk salient.

On the face of it, numbers alone would seem to be enough to deal with the Germans, so why not just maintain the very high level of T-34/76 production? It was not as simple as that. Even if it was the case, as the Soviets were now convinced, that the balance of the war in the East had shifted inexorably in their favour, the price the Red Army would have to pay in blood and treasure to achieve victory would surely increase the more they pushed the Germans back towards the Reich. Although the great expansion in weapons production in Germany, overseen by the Armaments Minister Albert Speer, had come too late to change the course of the war, not only was the number of panzers leaving the factories on the rise, but the qualitative effectiveness of those same machines was also increasing. While the Panther was deemed to have been a great disappointment in its combat debut at Kursk, it was still, according to German records, responsible for the destruction of more

Soviet tanks than any other type in that battle. Production of this formidable new machine was ramping up, and with more Army and SS formations being converted to operate it, the Panther would thus be encountered far more frequently and over the whole of the Eastern Front. Production of the Tiger I was also increasing and the Soviets had no doubt whatsoever as to the formidable armour and firepower combination of this heavy tank. Indeed, by the time of Kursk the Tiger was a known quantity to the Red Army with the first example having been captured in January 1943 near Leningrad. Examination of the new enemy heavy tank at the testing grounds at Kubinka provided the intelligence on its armour protection prompting the GKO on 15 April to issue an order to the People's Commissariat for Armament to develop new main guns to deal with these new German designs. To that end, at the close of April the Tiger had been subjected to firing tests employing many weapons in the Red Army artillery arsenal with a view to ascertaining a weapon that could fulfil that purpose.

It is worth noting, however, that at this stage, there was no intention of the chosen weapon being fitted to the T-34. Rather, these trials were being carried out with a view to finding a new main gun for the new heavy tanks under development. Of the weapons tested on the Tiger, the most effective was revealed to be the 85mm 52-K anti-aircraft gun. Of a calibre and performance similar to that of the 88mm Flak gun, it showed itself able to penetrate the 100mm-thick frontal armour of the German heavy tank at a range of 1,000m. In consequence, the 5 May GKO resolution 'On Enhancing Tank and Self-Propelled Gun Armament' directed the design bureaus of the Central Artillery Design Bureau, led by V.G. Grabin, and that of Armament Plant 9, led by F.F. Petrov, to develop two weapons each employing the K-52 anti-aircraft gun as a starting point, with the primary concern being to improve its ballistic properties. Two guns from each team were ready for testing by the following month, the speed of their development enhanced by the use of pre-existing elements from other weapons. That of Grabin's team, designated the S-53, utilised the cradle of the

76mm Zis-5 tank gun, while Petrov's team had employed the D-5S weapon as the basis for their submission, which they had designed for the SU-85 tank destroyer, shortly to go into production at URALMASH. Designated the D-5T, the advantage of the Petrov gun was that it had a low weight and its recoil was short – two qualities that would recommend its use in the necessarily confined space of a tank turret. Ultimately it was this weapon that was employed in the new heavy tank chosen for production which was ordered in September 1943 as the IS-85.

Post-Kursk, the acknowledged need to improve the firepower of the T-34 became a matter of profound urgency. This was no doubt reinforced by reports from participants in the battle that drew attention to the growing technical inadequacy of the T-34/76 relative to the German tanks they had encountered. General Pavel Rotmistrov, whose 5th Guards Tank Army had been devastated on 12 July at Prokhorovka, had to suffer the indignity of being subjected to a tongue-lashing over the phone from Stalin who demanded an explanation for what had been done 'to his magnificent Tank Army'. Nor was Rotmistrov prepared to suffer in silence over the matter. While unprepared to face Stalin and say what needed to be said to him (for obvious reasons) he turned to Zhukov as the Deputy Supreme Commander of the Red Army and demanded that the armoured forces be given a 'longer arm' to permit them to deal more equally with the new German panzers and upgraded older models.

ABOVE One of the perennial complaints of T-34 commanders was that their situational awareness was severely inhibited by the lack of a cupola with vision blocks. Although planned for on the T-34(M) of 1941, it took until mid-1943 for this feature to be added to production T-34/76s. The example here, on a drop-forged *Gaika* turret produced by URALMASH, is seen on one of the specialised variants of the T-34. The main hull machine gun has been replaced by a flamethrower. In this form the tank was designated as the OT-34. *(Copyright unknown)*

ABOVE The SU-122 was the first attempt to build a self-propelled gun on the T-34 chassis. It employed a modified M-30 122mm howitzer and was in production for only a short time. *(All on this spread Gennady Petrov)*

ABOVE An SU-122 in the finishing shop at Uralmash in 1942. It was common for workers and tank crews to daub patriotic slogans on their machines.

BELOW The SU-85 entered service with the Red Army in the later summer/early autumn of 1943. Armed with an 85mm gun its purpose was to provide heavy firepower support to the medium T-34/76 in dealing with new German tanks such as the Panther.

SELF-PROPELLED GUNS ON THE T-34 CHASSIS

During the course of the war three self-propelled guns were constructed employing the T-34 chassis. All three were medium SP guns as they fell between 20 and 40 tons in weight. They were an intermediate class between the light SU-76 and the heavy SP guns mounted on the KV and IS series chassis such as the SU-152 (1943), built on the KV chassis, and the ISU-122 and ISU-152 (both appearing in 1944). These were built on the IS chassis. The Russian initials SU stand for *Samokhodnaya Ustanovka*, meaning self-propelled gun. In the case of the three SU types described below, the casemate superstructure mounting the main gun was placed over that section in the T-34 that corresponded to the fighting compartment on the standard tank, that is, forward of the engine bulkhead.

SU-122

Originally designated the SU-35, this machine was not a product of the design team at Zavod No. 183 – the primary centre for T-34 production at Nizhne Tagil – but of the UZTM or URALMASH works. Under the design leadership of Z. Kotin, a Model 1938 122mm howitzer was mounted in a fixed, closed armoured superstructure mounted on a T-34/76 chassis. The armour was well sloped and the machine was thus well protected. It was the calibre of the weapon that prompted the change of designation to that of SU-122. Accepted by the GKO for production, the first machine was manufactured in December 1942 at URALMASH. First seeing service in January 1943, with the establishment of the first regiment of SP Artillery, the SU-122s were organised into platoons each of three machines where their main role was to provide artillery fire support, to tank divisions. It was assumed that the SU-122 would demonstrate the same versatility in combat as the German StuG III which had impressed the Russians with its ability to take on tanks when needed. The SU-122, however, did not live up to these expectations primarily because its high-explosive anti-tank (HEAT) round proved less effective than hoped. In consequence, its

production and service life was short. In a bid to rectify the anti-tank limitations, a prototype equipped with a longer-barrelled 122mm weapon was built. Although this imparted a higher muzzle velocity, the gun proved too heavy for the chassis and the project was cancelled, as indeed was the SU-122, after a relatively short production run. It was withdrawn from service in the autumn of 1943.

SU-85

First entering service in August 1943, the SU-85 was manufactured as a specialist tank destroyer. Its design was prompted by the need for the Red Army to field a more heavily armed machine than the T-34 with its 76mm gun, to assist the armoured formations cope with the new German Tiger I heavy tank which appeared towards the end of 1942 and the new Panther medium tank of which Soviet intelligence had gained knowledge before its combat debut at Kursk in July 1943. Inheriting essentially the same fixed armoured superstructure as its forebear, the heart of the SU-85 was its new gun. The 85mm D-5 was designed by F.F. Petrov and was derived from an anti-aircraft gun of the same calibre modified to be employed as an anti-tank weapon. That, mounted on the new tank destroyer, was designated as the D-5S – the 'S' standing for self-propelled.

Production began at the URALMASH Zavod in mid-1943, making it too late to be employed in the Battle of Kursk. The first units equipped with this new machine went into action in the Soviet offensives towards the Dnieper and into western Ukraine in the months thereafter.

Its ability to defeat a Tiger at 1,000m and to do equal damage to the Panther was greatly valued. During a production run of 2,050 machines running through to late 1944, the primary change which marked out the two main models was that in the initial variant the commander was given a hatch, but this was subsequently replaced by the same cupola as that fitted to the late versions of the T-34/76 – being designated the SU-85M. Another feature introduced towards the end of its production life and which marks it out as a late variant was a ball gun mantlet. Other improvements incorporated during its production life included better optics. What was never changed was the addition of a defensive machine gun for close-in protection – a characteristic of all the T-34-based SU machines.

What doomed the production life of the SU-85 was not its lack of effectiveness but the emergence in early 1944 of the T-34/85 which

LEFT Two German soldiers inspect a knocked out SU-85 in 1944.

BELOW The production life of the SU-85, like that of its predecessor, was also comparatively short by virtue of the fact that with the introduction of the T-34/85 in the spring of 1944 mounting the same calibre gun as the SU-85 there was no longer any value in retaining it. The army now wanted an SU with a bigger gun.

ABOVE A bigger gun came in the form of the new SU-100 mounting the D-10S 100mm weapon. The prototype drew heavily on the design of the SU-85, seen here on the right of the image. The final design saw a cupola installed for the commander on the left-hand side of the superstructure.
(All on this spread Gennady Petrov)

mounted the same calibre weapon. Although it was phased out of production to be replaced by its successor, surviving SU-85s received a new lease of life when they were re-equipped with the improved 85mm D5-S85A (M-1944) anti-tank gun and passed on to non-Soviet units' satellite forces, such as Polish forces serving in the Red Army.

BELOW Production of the SU-100 began at UTZM (Uralmash) in September 1944. The SU-100 saw service post-war in the Middle East.

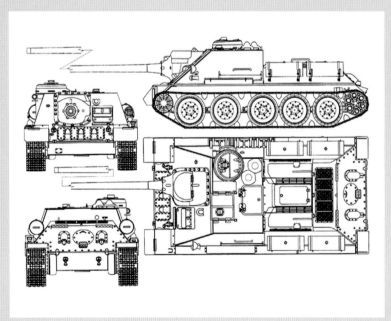

SU-100

The T-34/85 was already in production when in February 1944 L.I. Gorlitsky, the chief of the medium self-propelled gun bureau, turned his attention and that of his design team to developing a replacement for the SU-85. What emerged was the SU-100, a formidable machine whose effectiveness was such that until 1957 it remained the standard support gun with mechanised and armoured divisions in the Red Army, when it was replaced by the ISU-122. It remained in use with other satellite armies until much later.

Although sharing the same casemate structure as its progenitor, the SU-100 did, however, have thicker armour with the glacis being increased from 45mm in the SU-85 to 75mm in the new design, conferring on it a greater armour thickness by virtue of the angle of the armour. Internal space was marginally improved as the tank commander was now accommodated in a small sponson projecting from the right side of the hull, which was surmounted by a cupola with five viewing slits adapted from that carried by the T-34/85. This was also necessary given the bigger breech of the 100mm gun. In addition, the crew of four was assisted by the provision of a second ventilator that helped cope with the greater volume of fumes emitted by the bigger gun. Otherwise, the hull, running gear and automotive components were the same as that employed on the T-34/85. Indeed, of the total number of parts used in the construction of the SU-100 72% were taken from the T-34, 7.5% from the SU-85, with the remaining 20.5% being designed anew (most of these were to do with the gun mountings). Unsurprisingly, the new design was heavier than its progenitor with the SU-85 weighing 29.6 tons and the SU-100 31.6 tonnes. And while this design marked the high-water mark of the employment of the Christie suspension in Soviet design, it was to emerge without question as the most effective tank destroyer used by the Red Army and possibly the best of the war, the Jagdpanther notwithstanding.

As with its predecessor, the new tank destroyer would also mount a newly developed gun. In this case, development of the weapon preceded the design of the SU-100, as the need for a new anti-gun for the Red Army was recognised in 1943. This was as a result of intelligence acquired about the German intention to field even more heavily armed and armoured tanks in the near future – such as the Tiger II. This new weapon was also adapted from a pre-existing design. That chosen for the new anti-tank gun was the 100mm B-34 naval gun which was modified for its new role as an anti-tank weapon. As a towed gun it was also used in service as a field gun and received the designation 100mm field gun M1944 (BS-3). It was this same weapon, suitably adapted for employment as the D-10S, that was mounted on the new SU-100 tank destroyer. Employing a rifled barrel of 56 calibres in length, it had a high muzzle velocity. In testing, the 100mm gun of the new SU was able to penetrate both the vertical frontal armour of the Tiger I at 2,000m and the sloped armour of the Panther from 1,500m. In common with the SU-85, each of the new tank destroyers was equipped with both an intercom set and a radio. The SU-100 could carry 33 main rounds – these comprising the BR-412 armour-piercing tracer round and the high-explosive RP-412 round. It would, in due course and in a modified form, find itself equipping the post-war T-54/55 series of medium tanks.

ABOVE The length of the 100mm gun can be clearly seen in this photograph of a unit of SU-100s getting ready for operations. None of the SU-series was equipped with a secondary machine gun for close protection – it was deemed unnecessary as the task of the tank destroyers was to function as an over-watch to the less heavily armed T-34/85s.

The design of the new tank destroyer was initiated at URALMASH under the leadership of L.I. Gorlitsky in February 1944 and once again, this served to illustrate the Soviet ability to produce effective designs in rapid time. The first prototype, designated 'Object 138', appeared just a month later. With the SU-85 being phased out of production in August 1944, manufacture of the SU-100s came on-line the following month. Between then and July 1945, 2,335 were manufactured. The first SU-100s entered service with the Red Army in November 1944 and, in the following month as more became available, three new self-propelled Guards Artillery Brigades were created to employ the new design. With an HQ establishment of three SU-76s, each brigade fielded 65 SU-100s and had a manpower complement of 1,492 troops.

The SU-100 was extensively employed in the fighting against the German Army in 1945, making its combat debut in Poland. However, its most significant commitment was in Hungary where the 4th SS Panzer Korps was defeated in its attempt to break through to Budapest in January. Two months later, in early March, SU-100s were employed in large numbers to help defeat the abortive German offensive around Lake Balaton. They were used in the capture of Berlin and Vienna before the war's end.

Continuing to be produced in the USSR after the war, by the time it was taken out of production in 1956, some 6,000 machines had been built (including those constructed during the war). It was also built under licence in Czechoslovakia where a further 1,400 were manufactured. The SU-100 became a staple of Soviet arms exports, especially to the Middle East, and was operated by armies worldwide, including Yugoslavia (where they were used in the 1990s civil war), Egypt, Angola and Cuba. It was also sold to Communist China serving in the PLA as well as the North Korean Army. Indeed, its longevity has proven remarkable. Video evidence exists showing an SU-100 being used in Ukraine in 2014 and another being destroyed by an anti-tank missile in the ongoing Yemeni civil war in 2015! Both can be seen serving alongside T-34/85s in these conflicts.

BEUTEPANZER

In the Foreign Vehicle Classification system used by the German Army, the T-34 received the following designation: **T-34–747 (r)**. Tanks received a number in the 700 series with the country of origin indicated by a letter in brackets, in this case an (r) for 'Russland'. Although many hundreds of T-34s fell into German hands, between 500 and 600 actually served as *Beutepanzers*. This was the formal term employed to cover captured war *matériel* employed in German military service. These served in a variety of roles in tank destroyer or medium tank formations. While the intention was to formalise the use of such machines, quite often captured vehicles were employed in an ad hoc fashion by the unit concerned until the tank broke down and could not be repaired, owing to lack of spares. What is provided here is a sample of images depicting T-34s used by German units with a few of the photos coming from cameras belonging to ordinary soldiers.

1 Dating from 1941, an 18-ton Famo half-track is being employed to recover a T-34/76 Model 1941 that is in apparently good condition. This is an example that may be destined to be returned to Germany for evaluation at Kummersdorf.
2 A T-34/76 probably dating to the second winter on the Eastern Front as the crew is decked out in winter garb in a manner not available in the winter of 1941/42. The T-34 was captured and put into service.
3 A *Beute* T-34/76 late-model 1943 with the commander's cupola. This was finally added to the *Gaika* turret of the T-34 towards the end of 1943. The road wheels are distinctive of those used by Plant No. 183.

4 Some care has been employed on the *Beute* T-34/76. It has been repainted in the *dunkelgelb* with green overspray characteristic of German tanks from 1943 onwards. The large *Balkenkreuz* on the turret was necessary to prevent destruction from friendly fire.

5 A 1941 image showing a German-crewed captured T-34/76 model. It is useful for showing not only how large the main turret hatch was but how little the tank commander could see around it.

6 Another one for the family album! German soldiers pose with a *Beute* T-34/76 Model 1943. Note the even larger *Balkenkreuz* on the turret.

7 A late-model T-34/76 – the commander's cupola dating it to late 1943 or early 1944. Note the empty frame straps on the rear of the hull side where the cylindrical extra fuel tanks would be mounted. It appears as though it is being employed to help recover a broken-down StuG III.

8 This image could have been taken either in Russia or Germany. In the background is a captured Matilda II tank supplied to the Russians by the British through Lend-Lease. A lot of these were captured by the Germans after the Second Battle of Kharkov in May 1942. The hexagonal turret on the T-34/76 with the road wheel arrangement was distinctive of the No. 183 factory. This was the first Zavod to produce this turret.

9 An image showing two *Beute* machines. An STZ-built T-34/76, sans turret, has been converted into a makeshift recovery vehicle, probably by an engineering unit. In the background is a British Austin truck, large numbers of which were captured in 1940 in the French Campaign and used extensively by the German Army.

10 A picture dating from July 1944 shows a *Beute* T-34/85 built at the No. 112 factory in Gorki with the distinctive U-shaped lifting hooks on the turret. The time has been found by its captors to spray the machine in the full three-colour German camouflage scheme. Few T-34/85s were captured at that stage of the conflict in the East such that they could then be employed as *Beutepanzers*.

(All on this spread Anderson)

ABOVE The T-43 was the first attempt to radically update the T-34 design in the face of new German armour that had come on the scene. Emphasis was on improving the armour protection and providing a new three-man turret with a cupola for the tank commander. *(All on this spread Gennady Petrov)*

THE T-43

The intention to produce the much-modified T-34(M) in 1941 indicated that the military authorities were mindful from the start that the standard T-34 had limitations that needed to be addressed. But as we have seen, the onset of the war saw the T-34(M), which had been designed for that purpose, set aside in favour of the mass production of the standard T-34. However, in June 1942, the State Defence Council returned to the theme of modernising the T-34 and issued an order to that effect to the Morozov Design Team at Zavod No. 183 in Nizhne Tagil.

Although the designation T-43 suggested a completely new design, it was in fact an extensively modernised T-34. As such it was to feature:

■ A five-gear transmission
■ A commander's cupola on the turret
■ A simplified hull design to enable the employment of automatic welding in construction
■ An increase in the capacity of the fuel tanks
■ A torsion bar suspension.

The last of these was abandoned as the sole T-43 built retained the Christie suspension and running gear and the complete lower hull of the T-34. This was, however, reduced in length, which resulted in the elimination of the gaps between first, second and third road wheels.

The most radical changes were directed towards the upper armoured hull and the turret. By reducing the angle of the side armour it was possible to widen the fighting compartment, thereby increasing the space inside for the crew. This also permitted a wider turret ring, this being enlarged to 1,600mm to enable a new and improved turret to be mounted. Although the turret was bigger, it did not serve to raise the height of the machine, which can be seen in the comparative photograph of a standard T-34 and the prototype T-43 alongside. This new turret provided more space for the two men operating within it, although the T-43 was equipped with the same 76.2mm F-34 main gun with its co-axial Degtyarev light machine

BELOW Other changes included the use of a torsion bar system, but the T-43 was still armed with the same 76mm F-34 weapon.

RIGHT In this three-view drawing it can also be seen how the driver's hatch has been shifted to the left side of the glacis and the machine-gun position deleted altogether.

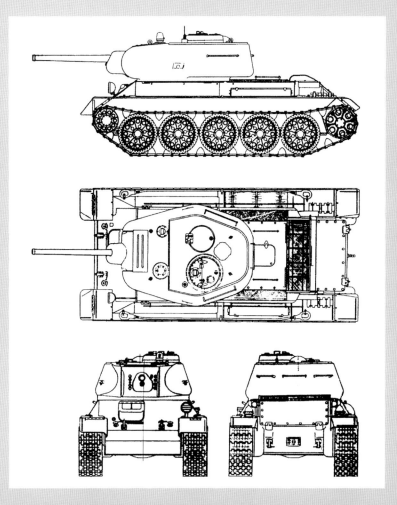

gun. The provision of a cupola for the tank commander finally dealt with one of the main criticisms of the standard T-34.

However, the glacis-mounted machine gun of the T-34 was eliminated and the glacis itself radically reworked with the armour being increased to 60mm and now set at an angle of between 52 and 55 degrees. Having eliminated the hull machine gun, the driver's hatch was now moved to occupy its place. The glacis was now only broken by the driver's hood. These changes led to a very different-looking machine, the dominating feature of which was the enlarged turret. As a consequence of these changes, the weight increased to 34 tons.

The sole T-43 was extensively tested in early 1943, performing well on trials, albeit not as well as the standard T-34. With the capture of the first German Tiger I heavy tank, and with Russian intelligence having got wind of information on the new Panther medium tank, it was recognised that what was really needed was not a tank with better armour, but heavier firepower. The T-43 was, in consequence, not accepted for production. The effort expended did not go to waste; the turret of the T-43 was adopted for the armament upgrade of the standard T-34 that emerged in late 1943 as the T-34/85.

BELOW This photograph graphically communicates the changes from the T-34 to the T-43. However, the new design did not go into production when it was accepted that what the T-34 needed was not heavier armour but a bigger gun. Suitably modified, the turret of the T-43 became the basis for that of the T-34/85.

ABOVE In a famous image dating from March 1944, a T-34/76 of the 1st Ukrainian Front carrying infantry 'tank riders' bypasses a burning Tiger I heavy tank as the Red Army pushed the Germans forces inexorably westward. *(RGAKFD Krasnogorsk via Stavka)*

BELOW Residing at the Finnish Army museum at Parola is an early T-34/76 Model 1941 captured by the Finnish Army from the Russians in the early period of the Continuation War (the term used by the Finns to describe their war with Russia, 1941–44, as an ally of Germany). As with serviceable tanks captured by them (or bought from Germany) they were modified to suit their own needs. This T-34 has been given headlights and covers for the same on the track guards and small storage boxes added. The road wheels are later types scavenged from T-34s knocked out later in the war. To the rear of the T-34 is British Charioteer tank destroyer supplied to the Finns in the 1950s. *(Jari Mäkiaho)*

So it was that in the months after Kursk, the Russians embarked on the search for a new gun for the T-34. This prompted Grabin to step forward and claim that the matter could be expedited by having his S-53 gun fitted into the standard turret of the T-34 Model 43. This did have a superficial attractiveness but, when attempted, trials quickly demonstrated that the gun was much too large for the existing turret. Even if the gun was right, it had to have a new turret. Zavod No. 112 in Gorki was given the job of overseeing this project where, under the leadership of V. Kerichev and his design team, they turned first to employing the turret of the now-defunct T-43, fitting it with an example of Grabin's weapon. Allocated the designation T-43-85, it was trialled but was rejected.

The turret of the T-43 had originally been designed to fit on to a modified T-34 hull wherein the engine deck had been lowered. As it would now be required to fit on to a standard T-34 hull with its raised engine deck, the turret would need to be modified to allow it to traverse without snagging. To that end, the turret was fitted with an armoured collar at its base to raise it, in order to permit full 360-degree traverse without impediment. There were also changes to the roof brought about by the need to accommodate the 85mm gun and its longer recoil. The gun needed to be slightly offset from the centreline in consequence of which the commander's position and cupola was moved from the centre of the turret and stationed behind the gunner. The gunner's hatch was deleted. These changes addressed the practical problems of adjusting the T-43 turret to fit the T-34 hull.

However, problems with the S-53 gun became apparent to Kerichev's team following their study of the first detailed technical drawings of the weapon, arising from its size relative to the space in the new turret. It was a little too large to permit the gun to be either properly raised or depressed. Grabin was thus ordered by Stalin and the GKO to resolve the matter, but in the meantime two T-34s with the new turret and the S-53 gun were despatched to Kubinka to undergo firing tests. These were successful enough for the GKO to order the tank into production. Nonetheless, with the new variant having been allocated the designation T-34/85 a number of plants began tooling-up to meet Stalin's injunction that it be ready for production by February 1944. In the meantime, testing of the S-53 gun continued, even though it too had been accepted for production. In late December 1943, however, a major flaw in the recoil system of the weapon was revealed which prompted a crisis in the production schedule. With a deadline for beginning manufacture just two months away, a desperate remedy was now required.

The solution was a two-stage response and was testimony to the rapidity with which the designers and industry in the Soviet Union could respond to an urgent challenge. Even as the search for a solution to the wider problems of the S-53 weapon began, to address the immediate contingency of meeting Stalin's production deadline, the Main Artillery Directorate (GAU) instructed Plant 112 at Gorki to temporarily substitute and equip the T-34/85s even now under construction with the well-proven and available 85mm D-5S weapon. So was born the first model of the T-34/85 – more often designated the Model 1943 – a product not of deliberate intention, but of expediency. In the meantime, the solution to the S-53's problems were surmounted by the creation of a new 85mm weapon that drew on the best aspects of four other 85mm weapon designs – the S-50 and 53 designed by Grabin, the D-5 from Petrov and finally the LB-85 – the latter being the product of a team working in one of the special camps in the Gulag! Given the designation Zis-S-53, this composite weapon was formally adopted by the GKO, displacing the S-53 as the preferred weapon for the T-34/85. Production of the new weapon began in early March 1944. By the end of that month it had begun to replace the D-5S on the T-34s manufactured at Plant 112 and was then taken up by all those others now geared up for production of the new variant of the T-34. By that date, the Gorki plant had produced approximately 800 of the tanks with the earlier weapon, and it was these that first saw service with the Red Army in the late winter of 1943/44. In no respect could the T-34 be seen as the ideal response to the new German technical challenge, but it nonetheless came at the opportune moment to assist in the huge Soviet offensives of 1944. To the Soviet military virtues of its availability in quantity was added its growing mechanical reliability and improved technology in the form of the new gun and three-man turret.

Into combat – the T-34/85 in 1944–45

This new version of the T-34 first saw service in March and April 1944, where it was employed by Katukov's 1st Tank Army. It sought to defeat a counter-offensive by the Germans as they attempted to break the encirclement of the 1st Panzer Army. Katukov was later to write of the new variant that:

> . . . The design of the T-34 with powerful new armament infused us with optimism and reinforced us psychologically. We could hardly wish for anything more when we saw that the new Soviet tanks [and he includes in this reference to the IS-II which also made its appearance for the first time in this operation] exceeded the often praised German tanks in combat capabilities.

The new gun, and more importantly the three-man turret, went a long way towards alleviating one of the principal design weaknesses of the 76mm-armed T-34s – that is the two-man turret. The addition of a cupola to the T-34/85 turret enabled a Russian tank commander to enhance his situational awareness by allowing him to place his head above it to obtain a

BELOW The first commitment of the T-34/85 to battle was in the spring of 1944 where the 1st and 2nd Ukrainian Fronts were attempting to effect the destruction of the German 1st Panzer Army. These Zavod No. 112-manufactured machines are serving with the 36th Guards Tank Brigade and are brand new. The new tank was greeted with great applause with special appreciation expressed by the Red Army tankers for its 'powerful new armament'.
(Gennady Petrov)

clearer view of his surroundings. When closed down, vision was aided by the provision of five fixed slits in the cupola wall and a Mark 4 periscopic sight mounted on the top of the cupola, in front of his hatch.

It was certainly the case that this conferred on the Soviet tank an undoubted technical superiority over the most commonly encountered German tank – the Panzer Mark IV. Even though this machine had progressively up-armoured and up-gunned after 1941, by 1944 the T-34/85 was superior to it technically, although other factors such as crew training and experience would more than likely determine the outcome of any confrontation between them. But once again the trump card for the Soviets was that they always had many more T-34s. The principal nemesis for T-34 crews was the Panther. The changes to the Soviet tank did not confer any degree of superiority over the German medium panzer which was increasingly being encountered even as the Red Army drove the German Army westwards throughout 1944.

One of the more interesting accounts of an encounter – more properly described as a duel – between a T-34/85 and a Panther was penned by V.P. Bryukhov and dates to 1944:

> . . . I personally knocked out nine tanks in my T-34/85 during the Yassy-Kishinev operation. I can remember one of the battles very well. We were driving through a cornfield, with corn as high as our tanks. Nothing could be seen, but the sorts of trails made by tanks were heading across the field in all directions. At a junction I saw a German tank drive quickly along a trail parallel to ours and disappear into the corn again (after the battle we found out it was a Panther). I ordered: 'Stop. Sight thirty to the right, tank at 400 metres.' Judging by the direction of his movement, we expected to see him again at the next crossing. The gun-layer traversed the gun to the right and we moved to the next trail. The German, in turn, had spotted us, and tried to bypass me through the cornfield. I looked into the panoramic sight at the place where he should emerge from the corn – and he did! We had to kill him instantly: if you let a German tank fire first and he missed with his first round, you had to bail out right away, as he'd always get you with the second one. German tank crews were like that. I shouted to the gun-layer 'Tank!' but he didn't see it. Half of the Panther's hull had already emerged from the corn. So I grabbed the gun-layer by his collar (he was sitting in front of me), threw him down onto the ammo storage, and took the seat myself. I aimed, and hit the Panther in the side. It caught fire like a petrol barrel and one of its crew bailed out.

Bryukhov's description is noteworthy for a number of reasons. There is an implicit respect for the effectiveness of German tank crews – indeed this section comes from a chapter entitled 'Only the luckier, smarter, sharper crews made it out alive'. He was clearly, luckier,

BELOW This Russian graphic shows the different models of the T-34/76 produced between 1940 and its withdrawal from production in 1944. The box on the extreme right gives the total number of that type produced. The T-34/76 was the most numerous type of T-34. *(Gennady Petrov)*

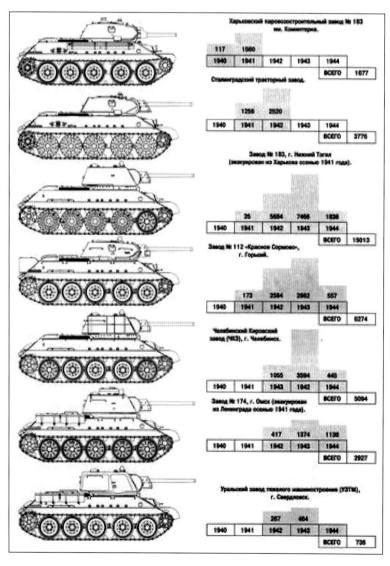

smarter and sharper knowing that he had to get the first shot in at the Panther and took matters into his own hands when his gunner was not quick enough to react. His observation that if they missed the first time, they wouldn't with the second, would see the crew bailing out before they did so, says much about the training and speed of response of German tank crews and the less effective training of the Russians. Whereas he might have had difficulty penetrating the frontal armour of the Panther, having it present itself side-on saw it at its most vulnerable, and his 85mm shell easily penetrated the 45mm flank armour. His description of the manner in which it 'caught fire like a petrol barrel' mirrors the experience of Allied tankers in Normandy. Even Heinz Guderian, the German Inspector of Tank Troops, commented on how easily the Panther burned when reporting back from the fighting in France. As with many tank crews in the Red Army, the longer they survived and garnered experience of fighting, the more effectively trained German tank crews, the greater the chances of their own survival. The corollary of that being, that many 'green' T-34 crews throughout the war rarely survived their first encounter, their demise and the destruction of their tanks contributing to the very high loss rates incurred in battle relative to that of the Germans.

It was also the case that tank crews found the engines fitted in their new T-34/85 charges to be more reliable than those of the earlier T-34/76s. This was reflected in the records of the testing of T-34s in the factory trials prior to releasing them for service, wherein they were driven to see what percentage of those tested could achieve a statutory minimum range of 330km without breaking down. The period covered begins in April 1943 and runs through to February 1944. Rather than detail all of the figures, we shall take those two months and September 1943, with them yielding percentage numbers of 10.1, 46.0, and finally in February 1944, 79.0. The first number is illustrative of the degree to which the quality of the engine build, even in the early months of 1943, was still very low, but how in the five months through to September this progressed markedly. By February 1944, the figure of 79.0 was a remarkable improvement. It pointed

to the general reliability of the T-34 improving substantially and indicated the far superior build conditions and engine component quality in the manufacturing factories. It also promised the T-34 being able to operate over longer ranges and this is perfectly illustrated by reference to the main Red Army summer offensive of 1944.

If the second half of 1943 had seen significant advances of the Red Army in the southern sector of the Eastern Front, the intention for their 1944 summer operations was to focus on the liberation of Belorussia

ABOVE T-34s were the cutting-edge of the Soviet counter-offensives post-Kursk and the offensives launched by the Red Army that continued all through the winter of 1943/44. *(Gennady Petrov)*

BELOW An early T-34/85 tank produced by Zavod No. 112 is taking cover to the side of a farm building during the summer of 1944. A distinguishing feature of these early D-5-equipped vehicles (when the mantlet could not be seen for identification purposes) was the four prominent lifting 'eyes' – two on each side – on the upper turret. Note the spare track link being carried on one of the turret grab handles. *(Gennady Petrov)*

ABOVE Operation Bagration, launched on 22 June 1944, marked the first occasion when the T34/85 was committed to battle en masse. Such was the rate of advance that fighting often continued into the night in this operation. *(Gennady Petrov)*

BELOW A column of T-34/85s of the 119th Rifle Tank Regiment mounting the D-5 gun built at the Krasnoe Sormovo Plant No. 112, spring 1944. Note the large drums of diesel fuel being carried on the rear deck. *(Gennady Petrov)*

and the destruction of Army Group Centre. The Germans were lulled by Soviet deception measures into believing that the main Red Army offensive would lay to the south of the Pripet Marshes and had in consequence deployed the bulk of their armour there. In practice this meant that of the 226 operational Panthers serving in the East in mid-June 1944, 175 were operating with Army Group North Ukraine and thus in the wrong place and many miles from where, on 22 June 1944, and the third anniversary of the German invasion, the Soviets launched Operation Bagration. The forces of four fronts, excluding the left wing of the 1st Belorussian Front – was fielding between them 4,070 tanks and self-propelled guns. Of the former, the majority was of T-34/85s with many of the self-propelled guns comprising SU-85s. The Red Army had been punctilious in ensuring that the bulk of T-34/85 production had been husbanded for this operation. Although the T-34/76 had remained in production alongside the T-34/85, of the 7,200 T-34s built by the beginning of June, no fewer than 6,000 were of the T-34/85 model. Three factories – namely UTZ No. 183, Krasnoe Sormovo No. 112 and Zavod No. 174 – were now responsible for T-34/85 production. Numbers of the T-34/76 were on the rapid decline with the last of this model leaving the production line of UTZM in March 1944. However, sufficient were still in service in May 1945 for numbers to be involved in the Battle for Berlin.

Here is not the place to examine Operation Bagration in minute detail. However, its execution led to the greatest defeat ever suffered by German arms conducted by a Red Army that operated in a fashion every bit as effective as that of the Germans when they had invaded Russia three years earlier. It also set the trend for the remaining offensives launched by the Red Army through to the end of the war, for all followed the example set by this operation in terms of the forces employed and distance travelled. This was the realisation of the concept of 'Deep Battle' as the vision had been dimly perceived by Tukhachevsky and his colleagues a decade earlier, and the presumption that it would hinge on the availability of 'masses of tanks'. Nor can there be any doubt about the centrality of the role of the T-34 in this massive operation. Having penetrated the forward German line, armoured units raced forward to reach the line of the River Berezina, 100 miles to the west. The encirclement and destruction of Army Group Centre, which was the primary objective of the first phase of the operation, was followed by a continuous pursuit of shattered and retreating units that had taken the foremost Soviet units as much as 300 miles to the west by early August, having liberated Belorussia and recovered much of Lithuania in the process. German losses attributable to the Russian offensive by that date have been reckoned at 277,000, with total losses for the whole of the Eastern Front by the end of August 1944

ABOVE The markings carried on the T-34 turrets enable these machines to be identified as belonging to the 26th Guards Tank Brigade, 2nd Guards Tank Corps, 3rd Belorussian Front in July 1944. *(Central Museum of the Armed Forces, Moscow, via Stavka)*

LEFT T-34/85s moving through a forest, summer 1944. Note the PPsH sub-machine gun carried on the front deck of the turret. At this stage of the war tankers were wary of German soldiers employing Panzerfaust anti-tank weapons in such close conditions. Where possible drivers would also keep their hatches open to keep the interior of the tank cool. *(Gennady Petrov)*

standing at 589,000. The foremost instrument in effecting this massive advance was the T-34/85 and the range of the advance telling testimony to the greater effectiveness of the engine that permitted such progress without the frequent breakdowns so common earlier on in its wartime career.

That said, it had still been very expensive in terms of tanks and SP guns lost, with as many as 3,000 written off in the course of the operation. This remains an extremely high number when it is considered that Army Group Centre was fielding just 553 AFVs, of which by far the bulk were assault guns. Many of the Russian AFVs that were lost had fallen victim to the increasing numbers of German defensive weapons. These included large numbers of the 75mm PaK 40 anti-tank gun, which was now standard equipment for the anti-tank troops, as well as the Panzerfaust ('Panzerfist', also called Faustpatrone meaning 'fist cartridge'). This was a one-shot throwaway weapon with a hollow charge warhead that could penetrate the frontal armour of any Red Army tank and was being issued in ever increasing numbers to German infantry. The problem for the infantry fielding

BELOW The replacement for the Tiger I in the form of the weightier but more heavily armed and armoured Tiger II first appeared on the Eastern Front in the late summer of 1944. The T-34/85, however, could get its measure as when on 13 August one belonging to the 53rd Guards Tank Brigade knocked out three in short order using the 'special' HVAP ammunition issued for such targets. As fewer than 500 were built before the end of the war and served on both the Eastern and Western Fronts, the small numbers encountered made little or no difference to the outcome in battling the Red Army, as its numerical superiority by this date was overwhelming. *(Author)*

LEFT A T-34/85 knocked out in the fighting in Rumania (as it was then known), minus its offside track, being examined by Hungarian troops. The sprayed-on star is a most unusual marking. *(Author)*

CENTRE The damage to the suspension on this T-34/85 is catastrophic. Not only has the heat of the fire burned off the rubber from the road wheels, but it has also caused the collapse of the large springs holding them up, Hungary, 1944. *(Author)*

them was that the range of these weapons was very limited, requiring them to have nerves of steel. The first variant, which went into production in October 1943, could penetrate 140mm of armour sloped at 30 degrees – but at a range of just 30m! Later models had increased range, but it is clear that they became real 'bogey weapons' for Soviet tankers (and also for Allied tankers in the West), no matter what tank they were in.

It is worth noting that with the close of 1944, total Soviet tank losses stood at 16,900 plus a further 6,800 SU self-propelled guns. Of the former number, over half constituted T-34s – indeed the percentage figure places it nearer 60%. These were by this date mainly T-34/85s but, as noted earlier, there were still substantial numbers of T-34/76s in service. The Soviet author G.F. Krivosheev observed when citing these figures that they constitute the largest number of AFV losses in a single year by any country in the history of warfare. By way of comparison, German AFV losses from a roughly equivalent period of December 1943 through to November 1944 totalled 10,070 machines. Even in the face of overwhelming and ever-growing numerical superiority, and the possession in the T-34/85 of a tank that was qualitatively at least equal to the later models of Pz.Kpfw IVH and J and StuG III, and nearly on a par with that of the Panther, these figures translate to three Russian

LEFT German soldiers swarm around a knocked-out T-34/85 in early winter 1944/45. The barrel of the 85mm gun is unusual in having a messy white line marking painted on top of it. *(Author)*

RIGHT As the Red Army drove into the Balkans and eastern Europe, the tank columns led by T-34/85s were at first greeted with genuine joy at being liberated from the Germans. Attitudes to 'liberation' were to change quite quickly. *(Author)*

CENTRE A T-34/85 knocked out in the late summer of 1944. Even equipped with the 85mm Zis gun, losses of the medium tank constituted some 58% of the total of 23,700 tracked AFVs lost by the Red Army in that year. *(Author)*

tanks destroyed for every one German, with most of these being T-34s.

As the purpose of this book is not to serve as a military history of the course of the war, operations through to the Fall of Berlin are not described. As 1945 opened, Germany's Reich was shrinking as pressure built from the West – where the British, Canadian and US armies were on the borders of Germany – and from the East as the Red Army built itself up for series of colossal offensives. By this date the very large number of T-34/85s being produced permitted almost all of the Red Army's tank corps to be re-equipped. These tank corps formed the spearheads of the massive winter offensives launched in early 1945. Within two months the German defences in East Prussia and Poland had been torn asunder and in a lightning advance, the Red Army arrived on the eastern banks of the River Oder, just 60km from Berlin, by the end of January. There then followed a halt, which saw the flanks being cleared. The fighting in Hungary saw the Russian forces contend with the last German offensive of the war around Lake Balaton, where ten panzer divisions – albeit many understrength – attacked the positions of the 3rd Ukrainian Front. The Russians absorbed the German assault and then counter-attacked, as did forces that had

RIGHT This T-34/76 Model 1943 with cupola carries the long beam attached to its glacis to which the PT-3 mine roller would then be attached when in use. It is passing two US-made GPA amphibious Jeeps supplied to the Red Army through Lend-Lease. *(Nik Cornish)*

ABOVE As it is possible to discern, albeit dimly, the driver of this T-34 behind his raised hatch, it must be assumed that at least one of the five men on the particular T-34 is not one of the crew! Clearly there is no expectation of combat being imminent as they all appear quite relaxed. *(Gennady Petrov)*

centre directly opposite Berlin (for it was to him that Stalin had initially given the honour of taking the German capital), and to the south, Koniev's 1st Ukrainian Front. Between them they were fielding some 6,200 tanks and self-propelled guns. When, on 16 April, the offensive began in the centre and south, Zhukov's forces got bogged down amid the German defences on the Seelow Heights and it took two days for the Russians to break through. A displeased Stalin erased the boundary lines between Zhukov and Koniev's Fronts, thereby opening up the capture of Berlin to either. That was enough to prompt Zhukov, but the delay at Seelow had already cost him over 1,000 tanks – the bulk of which were T-34s, many of which had succumbed to massed Panzerfaust fire. When finally they reached the city and began to advance into its environs, the T-34/85-equipped 11th Tank Corps had fitted screens made from bedsprings taken from German homes en route. These had been welded to hull sides and turrets as a form of improvised stand-off armour, the purpose of which was to cause the warhead of the German weapon to explode on the metal springs, and not on the hull of the tank. Many tanks, T-34s and IS-2s and SUs were to fall victim to this weapon. Although the following description concerns the impact of a Panzerfaust on an IS-2 heavy tank in the fighting for Berlin, it nonetheless is applicable to the T-34, as there were so many more of them than the heavy tanks.

BELOW By the spring of 1945, tank columns like this had reached the River Oder, just 60km from Berlin. These late-model T-34/85s were built at the No. 183 plant and are equipped with the Bsh smoke generators – these being mounted on the upper rear hull plate. *(RGAKFD Krasnogorsk via Stavka)*

been assembled for a pre-planned offensive on 16 March. Here the assault was led by the T-34s of the 6th Guards Tank Army – the same formation that had fought at Prokhorovka in July 1943. Subsequently these same forces would force the advance on and capture of Vienna which fell to the T-34s of that same formation on 13 April 1945. In all of these operations extremely large numbers of T-34s were involved.

The final assault on Berlin was launched just two days after the Austrian capital fell. Three fronts – Rokossovksy's 2nd Belorussian Front in the north, Zhukov's 1st Belorussian Front in the

Here is a tank with its hatches battened down. One can hear the whining of the dynamotor of its radio but the crew neither responds to banging on the armour, nor to radio calls. There is a small, penny-sized melt hole in the turret. It is even too small for a finger to get into it. A Faust's job. The skirt here was torn off and the shaped charge hit the armour. The welding torch is spitting bluish fire. Only the torch can open a hatch locked from inside. The bodies of the four dead crewmen are being taken out of the turret. Young lads. They might have lived on. The shaped charge burnt through the steel armour, bursting into the tank as a fiery whirlwind. Molten steel droplets killed everyone. Neither the stowage, nor fuel cells, nor mechanism were damaged, the people alone died.

RIGHT The build-up to the final assault on Berlin saw the massive stockpiling of equipment in the forests to the east of the River Oder. A T-34/85, carrying an infrequently seen small red star as part of its turret markings, passes an ISU-122 heavy assault gun in the foreground. *(Gennady Petrov)*

The fighting in the city was block by block as artillery and armour was used to blast German defenders from the buildings they had turned into defensive strong points. When finally the city was taken and the Germans surrendered on 2 May, the operation had cost the Red Army 2,156 tanks and SP guns, with the bulk of those, by virtue of their sheer number, T-34s.

There was a short pause as the Red Army savoured its triumph with the great Victory Parade in Red Square being led by the mass ranks of hundreds of T-34s. However, the T-34's war was not yet over. Hundreds were rapidly transferred to Mongolia as Stalin moved to honour his commitment to the Western Allies that Russia would enter the war against Japan not later than three months after the fall of Germany. And so it was that on 9 August Russian forces crossed the border into Manchuria and in a lightning campaign defeated the Japanese Kwantung Army. The

CENTRE The final offensive to take Berlin began on 16 April 1945. Fighting was extremely heavy – it had cost Zhukov's 1st Belorussian Front at least 30,000 men breaching the enemy defences on the Seelow Heights on the Oder itself. A race now began between Zhukov's and Koniev's commands – as Stalin had intended – as to which would take the laurels for conquering the city. In the three-week operation the Red Army lost approximately a third of all the tanks committed to the offensive. Many of these were T-34/76s, but many more were 85s. *(Gennady Petrov)*

RIGHT Surrendering German soldiers pass a late-model T-34/85 equipped with smoke generators. This machine was built at Plant No. 183 and was produced from the autumn of 1944. Although they cannot be seen clearly, it also employs the 'spider' road wheels – a distinguishing feature of T-34s produced at this Zavod. *(Gennady Petrov)*

LEFT Another example of a T-34/85 from the Nizhne Tagil Zavod. The more relaxed aspect of the soldiers and of the tank crew suggests that the fighting has either moved on or that this is an early post-battle image.
(All this spread Gennady Petrov)

CENTRE A T-34/85 moves slowly down one of the few less-damaged roads in the city. The large houses on the right of the image survived the Allied bombing campaign and the final assault on it by the Red Army. The burning wreckage in the foreground suggests that the fighting here is still continuing, hence the closed driver's hatch.

small number of enemy light tanks possessed by the Japanese were no match for the massed T-34s of Kravchenko's 6th Guards Tank Army. The campaign lasted some three weeks – and even though Japan surrendered unconditionally on 14 August – it did not suit the Russian forces to end the fighting until the 27th of that month. The cost in armour lost was fewer than 100 tanks. However, unbeknown to the Russian tank troops who had moved into the Korean Peninsula, their meeting with US troops on the 38th Parallel was to be portentous. Barely five years later, the T-34 would once again see combat, fighting in Korea against its one-time wartime ally.

BELOW With the iconic image of Berlin's Brandenburg Gate figuring in the background, Soviet troops savour their victory. These two famous images showing T-34/85s of the 36th Tank Brigade, 11th Tank Corps, are notable for the manner in which their crews have created impromptu stand-off defences made from bedsprings taken from German homes. These helped to prematurely detonate the feared Panzerfaust, which accounted for many Russian tanks of all types in the fighting.

BELOW Soviet and US troops first met at Torgau on the Elbe, thereby formally marking the division of Germany. A solitary US soldier sits atop the turret of a T-34, while a group of bemused but happy Russian troops pose for the picture.

RIGHT In a victory parade in Leningrad two T-34/76s with the *Gaika* turret roll across the great square in front of the former Winter Palace of the tsars. The one in the foreground still has AOHA over its glacis, suggesting that it is an earlier chassis with a later turret added during its rebuild. The column in the background celebrates another victory – that over Napoleon in 1812.

ABOVE AND LEFT When the time came for the great victory parade in Red Square, it was fitting that hundreds of T-34s trundled across its cobblestones. From first to last, Koshin and Morozov's creation had seen the Red Army through its darkest days to eventual victory. But the price for this machine was incredible. It is estimated that of the nearly 60,000 constructed in the war, some 50,000 were lost.

RIGHT The war against Germany was won, but the T-34/85 saw its final fighting of the Great Patriotic War (as the Second World War is known in Russia) thousands of miles from Europe in Manchuria against the Japanese in August 1945. These T-34/85s were constructed at Plant No. 183 at Nizhne Tagil. The next time the T-34 would see combat would be in Asia just five years later, against its former ally, the USA and its allies, in the Korean War.

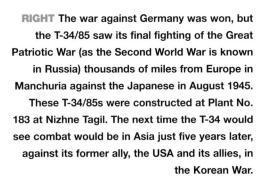

Chapter Three

Operating the T-34

Designed for mass production and manufactured in the most gruelling of circumstances, little concern was given to the needs of the crews of the T-34. Despite the interior being cramped and austere the crews adapted and praised the machine's main gun and armour protection, believing their tanks to be better than those of the enemy.

OPPOSITE Mikhail Smirnov, T-34 tank driver. *(Copyright unknown)*

For the Red Army crewmen who manned the T-34, more especially in the case of the 76mm-armed variant, the experience was one of mixed impressions. And indeed, what is offered here is at best a glimpse of what were perhaps the salient examples of their collective experience. While, as we have already noted, the basic design of the tank was very good, its potential, particularly in the first two years of the conflict, remained unrealised by virtue of flaws in other aspects of its design. Compounding these limitations were other factors such as poor crew training, which we have already discovered. When set against the backdrop of the flawed employment of the Soviet tank forces, which only began to abate from Stalingrad onward, it is possible to appreciate how in the years between 1941 and 1943 when T-34 production amounted to some 31,165 machines, no fewer than 23,600 of these were lost.

Ergonomics – that is, human factors engineering – was not deemed to be a priority in the original design of the T-34. Tankman Semyon Aria, in his recollection of service in a T-34/76 observed that 'care for the crew was limited to the most primitive things' and how bruises were unavoidable when driving and stopping. The padded leather headgear, known to the crews as *tankoshlems*, and still employed to this day in modernised form, were essential to protect heads from connection with the many items of equipment inside the hull. Another tanker expressed both envy and surprise at examining the interiors of Lend-Lease tanks.

I had a look at an American M4A2 Sherman [author's note: the variant supplied to the Russians]. My God! It was like a hotel inside! It was all lined with leather so that you didn't smash your head!

Unlike the crews of M4s, who could, at a pinch, sleep within their tank, it was simply impossible for Soviet T-34 crews to do the same. At the end of each day a makeshift tent would have to be created using the tarpaulin issued to them when they were given their tank. In consequence this item acquired great importance and its loss was not appreciated!

Many T-34 tank commanders stressed the importance of their driver; indeed, going so far as to suggest that their survival was heavily dependent on his expertise. But even then, the driver of a T-34 had a rudimentary metal seat that the Aberdeen Proving Ground evaluation team noted was fixed at a poor angle. Indeed, if driving for any length of time, the driver was so stiff that he had to be helped out of his position through his hatch in the glacis. Nor was his ability to drive his charge helped by the extremely problematic nature of the gearbox. Petr Kirichenko, who served as a radio operator/machine gunner, recalled how he and the loader assisted the driver in everything. When serving in the T-34/76:

[the] four-speed transmission took enormous effort to change the gears. The driver would set the lever in the right position and start to pull it, and I had to grasp the lever and pull it as well. It used to take several seconds. We had to perform this operation all the time during a march, and it was exhausting. During a prolonged march the driver usually lost two or three kilos.

Another driver, A.V. Maryevski recalled that 'it was impossible to shift the gear lever with one hand and I had to help myself with my knee'. Such was the consequence of the T-34/76 being required to operate with what was labelled the 'most primitive transmission of the time'.

For the radio operator in the first two years of the conflict it was the radio itself that caused great difficulties – if his machine was even equipped with one. As with other problems in the T-34 caused by the generally low level of technology, in the early days it utilised the 71-TK-3 set and it was noted that its capabilities were limited. It had a maximum range of about 6km when on the march, but it was deemed to be complex and 'unstable two-way'. It was not reliable, frequently broke down and was hard to repair. In 1942, such was the problematic nature of supply that equipping T-34s with radios virtually ceased. Obviously this impacted on the tactics employed by T-34 units in battle. In the absence of radios, unit commanders had little recourse but to use signal flags leading and employing what were designated as 'linear tactics'. In essence, the commander in his tank issued orders via flags to those tanks following him 'in line', a matter not assisted by the extremely large and heavy turret hatch fitted to the first generation of T-34/76s. This 'follow-my-leader' approach necessarily limited the ability of the T-34s to respond to German tanks, which by virtue of their far more effective command and control, were able to more rapidly manoeuvre against their Russian counterparts. From their perspective this led to 'T-34s operating in a disorganised fashion with little co-ordination, or else tended to clump together like a hen with its chicks'. Matters improved with deliveries and a better set being fitted once the production hiatus caused by the evacuation of the factories that produced them came to an end in the latter part of 1942. This improved model, designated the 71-TK-9R, proved to be far more reliable. In the T-34/85 the radio was shifted from the right-hand side of the hull next to the machine gunner/radio operator in the T-34/76, into the turret and was operated by the tank commander. The weakness in the radios was also reflected in the unreliability of the intercom system used between the commander in the turret and the driver who sat below and before him in the hull. To compensate, commander and driver found other means to communicate intentions with the former placing his boots on either shoulder of the driver. Poor communication became less of a problem on the T-34/85 with the introduction of the superior TPU-3bis intercom which was far more reliable.

The constrained space in the two-man turret of the T-34/76 also profoundly impacted on the ability of the tank commander to function effectively. V.I. Bryukhov described how, 'if you were a commander of a T-34/76, you had to do everything yourself – fire the main gun, lead the platoon over the radio, everything'. Lacking a cupola, the T-34/76 commander was normally unable to put his head outside the turret and so take in the situation – but to permit a degree of such vision and also to allow a quick release of the large and heavy one-piece turret hatch, some would leave it slightly open. Indeed, this may have been the difference between life and death if it came to having to abandon the tank. A.V. Bodnar said how, when going into battle, he would close the hatch but not latch it. He recalled how he would:

ABOVE Crew of T-34 tank named 'Suvorov', an early 112 Plant production tank. *(Copyright unknown)*

. . . hitch one end of his belt to the hatch latch and would wind the other end around the hook which held shells on the turret, so that if something happened I'd hit it with my head to make my belt come off and could jump straight out.

Preparation for such a contingency was very necessary for the forward view through the observation devices provided for the commander and loader was exceptionally poor – the Germans on examining the T-34/76 deemed it 'blind' for this reason. Bryukhov expressed a common view when he stated that Russian tank crews always noted the high quality of the Zeiss optics in the German panzers when compared

ABOVE **T-34/85 crew with four other tankers.** *(Copyright unknown)*

to their own, which, in the early T-34s, were polished steel mirrors being employed in both the commander's and the loader's observation devices. The Germans were quick to take note how this resulted in T-34 tank commanders having very poor situational awareness owing to the inferior quality of vision devices and a preoccupation with gunnery duties. Thus, in practice, the conflation of all of these elements resulted in a T-34/76 platoon (three tanks) being 'seldom capable of engaging three separate targets, but would tend to focus on a single target selected by the platoon leader. As a result T-34 platoons lost the greater firepower of three independently operating tanks.' It also served to reduce the volume of the firepower from the said three tanks because of their difficulty in finding and engaging targets. In consequence, the Germans could exploit this limitation, and notwithstanding their inferior main armament, get off three rounds to each one from the T-34. It was only in the later models of the T-34/76 1943 model with the *Gaika* turret that tank commanders were finally given the cupola they had been asking for. It was of course, standard fitting on the T-34/85.

For the crew in the T-34/76 one of the main problems that emerged every time they went into combat between 1941 and the introduction of the *Gaika* turret in mid-1942 (which sought to address this problem) was the build-up of fumes from the firing of the F-34 main gun. The lack of an effective extraction fan saw these pervade the whole of the fighting compartment, although it was the loader who seems to have suffered the most. Bryukhov recalled:

You'd yell: 'Load anti-armour, load anti-personnel', then you'd turn around and see the loader laying lights-out on the ammo boxes [on the floor]. He'd been poisoned by fumes and lost consciousness. Few could last out a heavy fight to the end.

A lack of forethought in the design process had seen the extraction fan located in the same position on the front of the T-34 turret as it had occupied on that of the BT-7. This corresponded to where within the turret the much shorter 45mm gun breech would have ended on the BT-7. However, that of the 76.2mm F-34 was much longer, projecting further into the body of the turret, so that when the gun was fired toxic fumes vented from the breech every time a shell case was ejected, which the fan, being so poorly positioned, was unable to fully extract.

The loader always had especial problems, be it in the 76 or the 85 model. Whenever the ammunition stored in the clips on the inside of the hull or turret had been used up, he would then need to reach down and retrieve rounds from the wooden boxes in which they were stored on the floor of the tank and on which the crew stood. In the heat of battle this meant leaning down in a very constricted space, moving aside the cover and opening up the wooden boxes in which the rounds were stored. He would need to know where to get the appropriate round signalled to him by the commander. This would then be unstrapped from the box, lifted out and then stuck in the breech. Lacking a turret basket, if the turret needed to be traversed as he was retrieving ammunition, the whole position would slew around and he would need to be agile enough to avoid being hit by the gun breech as the turret moved. Furthermore, the sheer physical effort required to lift and load the shells one after the other often found the loader as physically spent as the driver. This was even more so in the later T-34 as the weight of the 85mm rounds was greater than that of the 76mm ammunition.

But perhaps the most difficult time for service in the T-34 was in the winter. Soviet tankers had no special adaptation to the privations of the Russian winter even though in that of 1941/42 they were more appropriately dressed than were their enemies for the extremes of the cold. The T-34 had no heater for warming the interior of the tank, so all of the crew had to wear the heavy winter sheepskins that bulked them up and made movement in an interior already very cramped even more so. This was especially true for the commander and the loader whose ability to move around in any measure was even more constrained. Nor was it the case that the engine was less prone to being affected by the extremes of cold when compared to German tanks. However, the greater familiarity with the extremes of the climate meant of course that Russian tankers were better prepared to know how to deal with them. It was known that it was difficult to start up the V-2 diesel engine in the winter and it had to be warmed up at least two hours before departure. Tank crews were told to employ a tray that would fit between the tracks, place it underneath the engine having filled it with diesel fuel and set it alight. Normally such a procedure required monitoring by the crew to prevent any mishaps. After about an hour and a half both tank and crew would be covered in smuts from the burning fuel. Messy, yes, but the engine of the tank would function even in extreme cold.

Tank crews of any nation are a singular breed. But there is no question that the crews of the Red Army serving in the T-34 between 1941 and 1945 possibly had one of the hardest wartime tasks of any. It was a machine designed to service the needs of the crews who manned it and of the culture that created it. Although a somewhat glib generalisation, it nonetheless sums up what so many Russian tankers still believe about this tank and the role it played in securing victory in the Great Patriotic War when they say that 'against the T-34 the German tanks were crap'.

BELOW T34/76 crew. *(Copyright unknown)*

Chapter Four

T-34s in post-war foreign service

Post-1945, the T-34 became a primary Soviet military export to allies and client states during the early Cold War. It saw combat in Korea, the Middle East and wars in Africa through to Vietnam. Its longevity has been such that a few are still in service in the Yemeni and Syrian civil wars.

OPPOSITE Young soldiers from the East German Army crowd around a T-34/85 in Rostock in April 1962. *(Bundesarchiv Bild 183-92246-0004)*

It has been many decades since the first T-34 Model 1940 left the production line at the Kharkov Plant No. 183. The design has been obsolescent for many years since then. However, T-34/85s are still being used in combat at the time of writing in the civil wars in Syria and Yemen. They are also still to be found serving in the armies of a number of African and Asian states, albeit most are now in reserve. This is a remarkable longevity for a tank design, comparable to considering the British Mark IV tank of 1918 still seeing service in 1995! Since 1945, the T-34/85 has been utilised by a very large number of armies. For the Soviet Union, the export of arms either was as a means to secure foreign currency – many countries purchasing them were required to pay in dollars – or as largesse, freely given to secure influence with 'fraternal' countries in the Cold War. The table at top right identifies all those countries that acquired T-34/85s to serve in their armies after 1945. A short summary is then given of the role of the T-34/85 in the many conflicts it has served in since 1945.

Countries which acquired the T-34/85	
Europe	Albania, Austria, Bosnia, Bulgaria, Czechoslovakia, East Germany, Greece, Hungary, Poland, Romania, Slovenia
Americas	Cuba
Middle East	Iran, Iraq, Lebanon, North Yemen, South Yemen, Syria
Asia	Afghanistan, Cambodia, Laos, Mongolia, North Korea, Pakistan, P.R. China
Africa	Algeria, Angola, Rep of Congo, Egypt, Equatorial Guinea, Ethiopia, Guinea, Guinea-Bissau, Libya, Mali, Mozambique, Namibia, Sahrawi Republic, Somaliland, Somalia, Sudan, Togo, Zimbabwe

The T-34/85 in combat: 1950–2017

The intention of this section is not to give a detailed account of the combat record of the T-34 in the post-war period but to offer a brief account of the main conflicts in which it has been employed. Of those covered below it was only in the first, the Korean War, that the T-34/85 was to play a key role. Thereafter, in the Middle East conflicts of 1956 and 1967, although serving in some numbers in Egypt and Syria, it increasingly played second-fiddle to the later T-54 as these began to be exported to those countries in its stead. The first recipients of the T-34/85 were those armies in Eastern europe that from 1955 onward were signatories of the Warsaw Pact. T-34s were quickly phased out as they were replaced by the T-54/T-55 series of medium tanks. T-34s did, however, see internal 'service' in East Germany and later Hungary in the suppression of popular uprisings in those countries in 1953 and 1956 respectively.

T-34s were produced under licence in Poland, Czechoslovakia and China. In the latter, the T-34/85 was known as the Type 58 but service with the People's Liberation Army (PLA) was limited by virtue of it being replaced on the production line by the licensed copy of the T-54, known in China as the Type 59. The Type 58 was phased out of PLA service by 1960.

BELOW **The T-34 served in the Red Army in extensive numbers in the immediate post-war period. This was primarily a consequence of the failure of its planned replacement, the T-44, to satisfy the Army's requirement for a new medium tank with a gun with a larger calibre than 85mm. This only came with the T-54 in the early 1950s. In the meantime, it also equipped the armies of the countries of eastern Europe in large numbers and was built under licence in both Poland and Czechoslovakia.** *(Gennady Petrov)*

LEFT A damaged T-34/85 left abandoned by its Red Army crew being examined by a number of inhabitants of Budapest in the very early stages of the 1956 Hungarian Uprising. *(Copyright unknown)*

The Korean War, 1950–53

The Stalin-sanctioned North Korean invasion of South Korea on 26 June 1950 was spearheaded by some 150 Russian-supplied T-34/85s of the North Korean People's Army (NKPA) constituting the 107th, 109th and the 208th Tank Training Regiments of the 105th Armoured Brigade. The remainder of the 242 T-34/85s supplied to the NKPA were kept back and dispersed among the five further tank regiments which were available to be used in follow-up operations, should they be needed. When the bulk of US forces left the country, with only small training units remaining, the South Korean Army had only been equipped by the Americans with light AFVs. This was owing to the belief that by providing national leader Syngman Rhee with tanks, he would be encouraged to invade the north to forcibly reunify the country. The irony could not have been lost when the NKPA did the same, but the other way.

The North Korean T-34s were expected to be the primary instrument to effect a total and rapid victory over the south. Kim Il-Sung convinced Stalin that now the Americans had left (and were unlikely to intervene) and with the South Korean military so poorly equipped, the operation would be wound up within a month from the start of the invasion. It was only then that Stalin gave the green light for the NKPA to proceed.

BELOW The Korean War was the last major conflict in which the T34 played a dominant role. Almost all of the T-34/85s supplied by the Russians to the NKPA had been destroyed by the time of the ceasefire in mid-1953. *(NARA)*

In the initial period of the war the T-34/85 swept all before it gaining a reputation for invincibility. But once the USA introduced heavier armour such as the M4A3E8, the M-26 and M-46, the Russian-supplied tank met its match and increasingly became the victim. It was the M4 that inflicted the greatest attrition on those T-34s that were lost to tank-versus-tank action. With total control of the air, it was also extremely difficult for NKPA armour to move in the daytime without being attacked. Although fighting its first war in Korea, the British Centurion had no encounters with T-34s. *(All this page NARA)*

Initially all went well, but having inaccurately gauged the US response, NKPA ground forces were very quickly targeted by effective US air power. The rapid reaction of the USA saw M-24 light tanks from the occupation force in Japan arriving in South Korea in early July with the first encounter with T-34s taking place as early as 10 July. This and others showed the M-24 was no match for the T-34 with just 2 of the 14 originally committed being left by the beginning of August. However, by that date, attrition – through combat losses and the inevitable breakdowns – had whittled the North Korean T-34 strength down to just 40 tanks.

Having withdrawn to the Pusan perimeter in the south of the country, the US and South Korean forces now awaited armour reinforcements which by the end of August included Army M4A3E8s and Marine M-26 Pershing medium tanks, and in a number of encounters the US Marine and Army tanks demonstrated their ascendency over the NKPA T-34s. On 17 August a well-laid ambush saw Marine M-26s destroy three T-34s. Hitherto regarded as unbeatable this encounter, according to the official US Marine history, served to shatter 'the myth of the invincible T-34 in five flaming minutes'. Ten days later M-26s and infantry stopped an attack by NKPA armour by T-34s supported by SU-76 light tank destroyers – and over two days' heavy fighting the NK forces lost 13 of the former and 5 of the latter. Daylight attacks became increasingly rare as US air power targeted any vehicle that could be seen moving – the T-34 being the primary target – with napalm, doing great damage to any AFV caught in the open. The last real attempt by the NKPA tank forces – and by this date the North Koreans were having to commit the less well-trained cadre tank units – took place in the opening days of September when a force of T-34s and infantry made last-gasp assaults on the Pusan perimeter and also later in September along the line of the Nakhtong River. By then the military situation had been transformed by the UN landings at Inchon, on the central west coast of Korea. This was many miles to the rear of the NKPA forces still fighting around Pusan, who had no choice but to withdraw the remains of their armoured forces northward to help contest the UN expeditionary force landings and its advance inland towards the South Korean capital, Seoul.

Although the NKPA committed their remaining T-34s in some numbers to try to halt the UN advance between the landings and 20 September, by that date many of the surviving T-34s had been knocked out by a combination of air power and the guns of mainly the Marine M-26s. The breakout at Pusan planned and executed on the day after the Inchon landings served to eliminate or capture those T-34s that had not moved northwards. By October, the first M-46s had entered combat in Korea with the 'blooding' of Company A of the 6th Tank Battalion, US Army, in an encounter with eight T-34s and one Su-76M, all of which were knocked out.

After October 1950, there were exceedingly few armour vs armour encounters. A subsequent post-combat assessment made in 1954 concluded that there had been a total of 119 tank vs tank encounters since July 1950, of which the majority were conducted by US Army units with the remainder by the US Marine Corps. When compared with the size of tank encounters in the Second World War, those in Korea were small, with it being noted that only 24 of those 119 engagements involved three or more NKPA tanks. For a loss of 34 tanks which could be attributed to T-34 or Su-76M fire (of which just 15 were total 'write offs'), the US armour knocked out 97 T-34s and a further 18 as probables. Of the total number of T-34s available to the NKPA at the start of the invasion, by November 1950 223 had been destroyed and 33 were damaged, giving a total loss to all causes of 256 machines.

The reputation garnered for 'invincibility' by the T-34/85 in the opening month of the Korean War was purely a consequence of the South Koreans having very few effective weapons to combat it. The first US tank to encounter it, the M-24 light tank, was not equipped to engage with a more heavily armed and armoured medium tank, so unsurprisingly came off second best when it came up against the T-34. Once, however, the M4A3E8s and M-26s started to arrive in the country, the T-34 was rapidly mastered. And after November 1950, with their tank force destroyed, the

OPPOSITE BOTTOM

Israeli troops inspect three SU-100s and a solitary T-34/85 abandoned somewhere in the Sinai Desert in the 1956 war. *(NARA)*

LEFT The T-34/85s supplied to Egypt by Czechoslovakia saw combat in both the wars of 1956 and 1967 against the Israeli Defence Forces, with many being lost in the fighting in the Sinai in both conflicts. By 1967 it had been largely relegated to an infantry-support role as its place as a main battle tank was taken by the T-54/55 in the Egyptian Army. *(Copyright unknown)*

character of the war changed. The intervention of the Chinese in large numbers did not see a revival of armoured warfare as they chose not to commit their T-34s to the conflict.

In terms of combat with the armoured forces of the major powers, it was clear from the Korean experience that the T-34/85 had had its day. A new generation of Allied armour, such as the US M-46, M-47 and M-48 and the British Centurion, saw the Soviet response in the form of the more heavily armoured and armed T-54 and T-55. The T-34 was initially employed to equip the resurrected armies of eastern European satellite states and even manufactured under licence for that purpose in Poland and Czechoslovakia, but even in the armies of those countries, the T-34 was eventually replaced by the more formidable T-54 from the late 1950s onward. However, Korea was not the only place to witness the T-34 in action in that decade.

The Arab–Israeli Conflicts, 1956 and 1967

T-34/85s and SU-100s were acquired by Egypt from Czechoslovakia in the period prior to the war of 1956, wherein the T-34s served mainly in the infantry-support role. Captured examples were displayed in post-war military parades by

CENTRE A T-34 knocked out by the Israeli Army on the Golan Heights in the Six-Day War of 1967. Note the machine gun ring fitted to the commander's cupola by the Syrians. *(Copyright unknown)*

LEFT A knocked-out Syrian SU-100 in the Valley of Tears, 1973. *(Copyright unknown)*

the Israeli Army. In the Six-Day War of 1967, many T-34s and SU-100s were knocked out and captured by the Israeli Army, but by this date the increasing obsolescence of the former saw them being 'dug in' and used by the Egyptian Army as fixed strongpoints. Thereafter, many of the surviving T-34s in Egyptian Army service were converted into self-propelled guns by the mounting of the Russian 100mm BS-3 anti-tank gun of 1944 vintage within a lightly armoured superstructure.

Russian-supplied T-34/85s were also employed by the Syrian Army where they were involved in gunnery duels with the Israeli Army along the Golan Heights before the 1967 war; many were knocked out and captured. Most Syrian T-34s were equipped with a heavy machine gun mounted on the turret roof.

In other theatres

Elsewhere, the T-34 saw combat in Cuba where it was employed to help defeat the Bay of Pigs Invasion in 1961, by the North Vietnamese Army in the Tet Offensive of 1968 and again four years later in the 1972 Easter Offensive in South Vietnam.

It was also utilised in the Angolan Civil War against the People's Movement for the Liberation of Angola (MPLA) and also in fighting with the South African Defence Forces in the 1970s. That same decade saw its employment by the Cypriot National Guard in combating the Turkish invasion forces of the island in 1974.

The Vietnamese Army was supplied with a large number of Type 58 tanks – the Chinese copy of the T-34/85 – and these saw employment in the Tet Offensive of 1968 and the Easter Offensive of 1972 where they proved vulnerable to the US and South Vietnamese Army M-48s and M-41s. In the final stages of the Vietnam War, the Type 58 saw much less action as the Chinese-manufactured and -supplied Type 59 bore the brunt of the armoured fighting. *(NARA)*

RIGHT The conflicts following the break-up of the state of Yugoslavia in the early 1990s saw the various parties involved in the fighting drawing on both contemporary and older stored equipment – in this case T-34s as many had been held as war reserves by the former Yugoslav Army. Operated by Bosnian Serb forces, this T-34 has been fitted with sheets of fibre/rubber matting in a possibly futile attempt to reduce its infra-red signature in the face of weapons designed to home in on such. *(Copyright unknown)*

RIGHT Both in the Yugoslav Civil War and latterly in the fighting in eastern Ukraine, T-34/85s have seen combat, albeit in much fewer numbers in the latter. This example has come to a sorry end, having broken its 85mm gun barrel in consequence of having driven off the road into a ditch or having been bulldozed there. Of note is the heavy machine-gun pillar that has been welded to one of the double mushroom vents on the rear of the turret's roof. The tank was probably produced in late 1944 at the Plant No. 112 in Gorki. *(Copyright unknown)*

The type also saw service with the Iraqi Army in the conflict with Iran.

In the period between 1990 and 2017, this venerable old warhorse has seen further action in the fighting that accompanied the break-up of Yugoslavia. The T-34, as with other types of older AFV, was heavily involved in the fighting especially in the early days where the combatants employed any fighting vehicle that lay to hand. More recently, the T-34/85 has

LEFT A picture taken in 2017! It shows a T-34/85 being employed in combat operations in the Yemeni Civil War, employing a camouflage scheme reminiscent of that used by the Germans from 1943 onwards. *(Yevgeny Scharov)*

seen service in the Syrian Civil War (this can be seen in film on YouTube), as well as that in Yemen, where both the tank and the SU-100 have seen combat in 2017.

Despite its age, if there is a T-34 to hand in these conflicts, it is serviceable and there is ammunition available, it is pressed into service. It would seem that we may yet catch a glimpse of the T-34 in action on our television screens for a few years yet.

LEFT From the same photographic source is an SU-100 driving through one of the main streets of the Yemeni capital Sana'a. With T-34s also being used in the fighting in Syria at the time of writing (there is film of this available on YouTube), it is a testimony to the longevity of this fighting machine whose origin dates back, as we have seen, to the late 1930s. *(Yevgeny Scharov)*

Chapter Five

Anatomy of the T-34/76

Produced from 1940 through until 1944 when it was replaced on the production lines by its more heavily armed successor, the T-34/76 was a revolutionary design when it first appeared. Although the hull changed little in appearance over that time, the terrible exigencies of war facing the Soviet Union wrought great internal changes on the interior to simply production.

OPPOSITE A T-34 Model 1943 with the 'Gaika' turret and commander's cupola is repaired in the field by a mobile workshop and its crew. The picture was taken in the summer of 1944 during Operation Bagration and the tank itself belongs to the 36th Guards Tank Brigade of the 4th Guards Mechanised Corps. *(Nik Cornish)*

Hull

The heading of this section recognises from the outset that while there was a basic hull configuration common to all T-34 tanks, beyond that there was a multiplicity of features, both internal and external, that identified the factory of origin of any particular tank. Many of the details of the tank both externally, and to a lesser extent internally, are covered by the 'walk-around' of the T-34/85 held in the Tank Museum at Bovington (see pages 125-129) supported by a few others from the T-34/85 held in the collection at the Royal Military College of Science at Shrivenham. Particular information on the detail of the different aspects of the tank is derived from the report on the evaluation of the T-34/76 examined at the British School of Tank Technology in 1943.

In terms of the machine at its most basic, the hull was a rigid armoured box constructed from rolled homogeneous armour plates welded throughout. The housings for such components as the final drives and the idlers were all welded to the hull. Examination of the accompanying diagram shows both the angles and thicknesses of the armour employed on the hull. Although that illustrated in the diagram is of a T-34/1940, apart from a different turret, the very last T-34/85s produced under licence in Poland and Czechoslovakia post-war employed a hull whose dimensions and specifications had changed hardly at all, other than that of the thickness of the side armour as described below. In such a long production run, it is surprising that, when compared to the German Panther (the machine ostensibly developed by the Germans to counter the T-34), the Russian design evidenced no substantive change to its appearance as did the Panther in the two years of its production run. The hull of the Panther Ausf D – the first model in production from January to September 1943 – looked different from that of the last variant, with the Ausf G having a redesigned hull. While the turret of the T-34/85 encountered by the US Marines and Army in Korea in the early 1950s clearly differed to that of the T-34 model 1942, the hull employed by both had not changed at all in its essential form.

BELOW One of the original blueprints for hull plan and profile views of T-34/76. The tall column within the hull are to house the spring coils of the suspension. *(Gennady Petrov)*

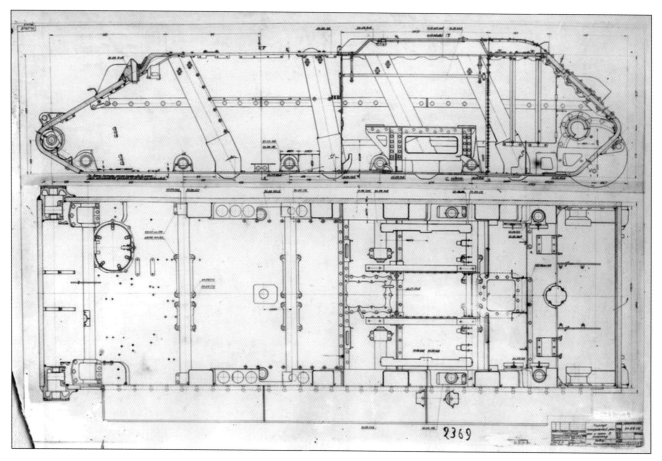

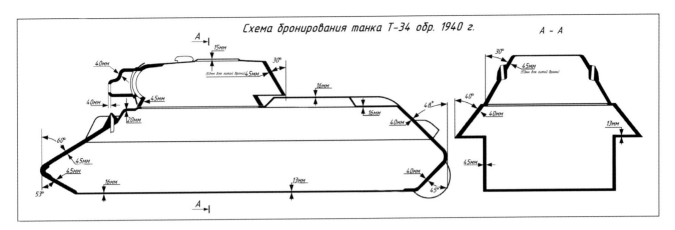

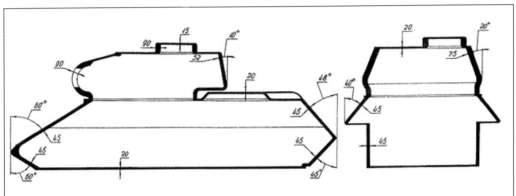

ABOVE AND LEFT The armour thicknesses for hull and turret of T-34/76 and T-34/85 provided for comparison. *(Gennady Petrov)*

The bottom plate was the weight-bearing element of the hull, with the front section being 3mm thicker than the rear. The front and rear parts, of 45mm and 40mm respectively, were connected to it by welding. This in turn was reinforced to ensure rigidity by a cast T-shaped beam which also formed part of the framework of the engine compartment bulkhead. This beam was both welded and riveted to the bottom plate on either side of the seam. Neither the front nor side armour of the T-34 was especially thick. That of the frontal glacis was 45mm sloped at a 60-degree angle from the vertical, which conferred upon it a thickness equivalence of 75mm (although other sources give a different figure). In the first production model of the T-34, as shown in the diagram, the thickness of the hull side armour was 40mm, increased to 45mm on the Model 1942, introduced at the end of that year. It would stay at 45mm until production of the T-34 ended with the angle of the side armour on all models being 40 degrees. The rear armour plate which gave access to the engine was understandably thinner at 40mm, set at an angle of 48 degrees, which while not as effective as that on the front or side, nonetheless provided protection to the transmission and the engine. This plate was hinged and fixed in place by a number of bolts. Before 1942 access to the transmission without removing the rear plate was possible by the provision of a rectangular hatch. This was replaced on all T-34s from 1942 onwards by a circular hatch, as can be seen in the images of the T-34/85 provided in the T-34 walk-around section. The only factory to retain the original rectangular hatch was STZ, but this only lasted until the Stalingrad factory ceased production towards the end of 1942. A characteristic and identifying feature in photographs of the T-34s produced by the STZ and adopted by the Krasnoi Sormovo factory in Gorki was a simplified hull design in which the glacis and rear hull plate was welded into interlocking sides. The armour on the top of the hull was noticeably thinner (as was/is universal on tank hulls of whatever nationality) and was engineered to house a turret ring on the Model 1940 through to the Model 1943 of 1,440mm. The need to take the bigger turret of the T-34/85, with its larger and more powerful gun and three-man crew, saw this widened to 1,600mm.

Cutaway of T-34/76 1943 model. *(Mark Rolfe)*

1 Radio aerial
2 Cast turret
3 Lifting eye (x4)
4 Gunner's hatch
5 Viewing periscope
6 Commander's cupola
7 Air louvres
8 Engine grille
9 Fuel tank
10 Idler wheel
11 Road wheel
12 Cast manganese steel tracks
13 Tool box
14 Horn
15 Headlight
16 Drive sprocket
17 Mudguard
18 Driver's seat
19 7.62mm DGT machine gun
20 Machine gun mantlet
21 Combined towing and lifting hooks
22 L11 76.2mm F-34 gun
23 Glacis plate

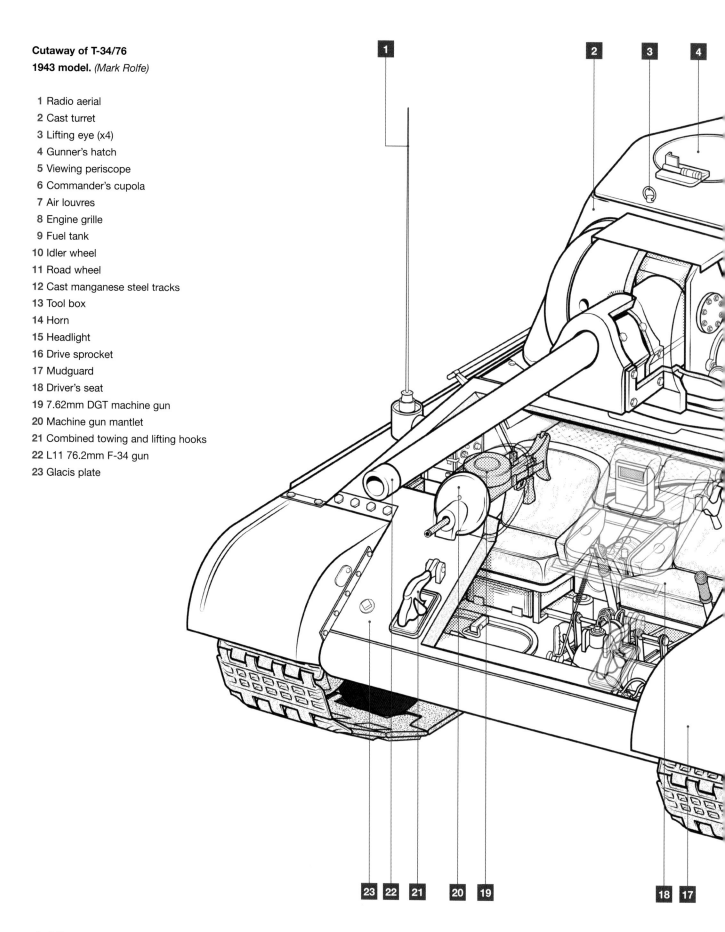

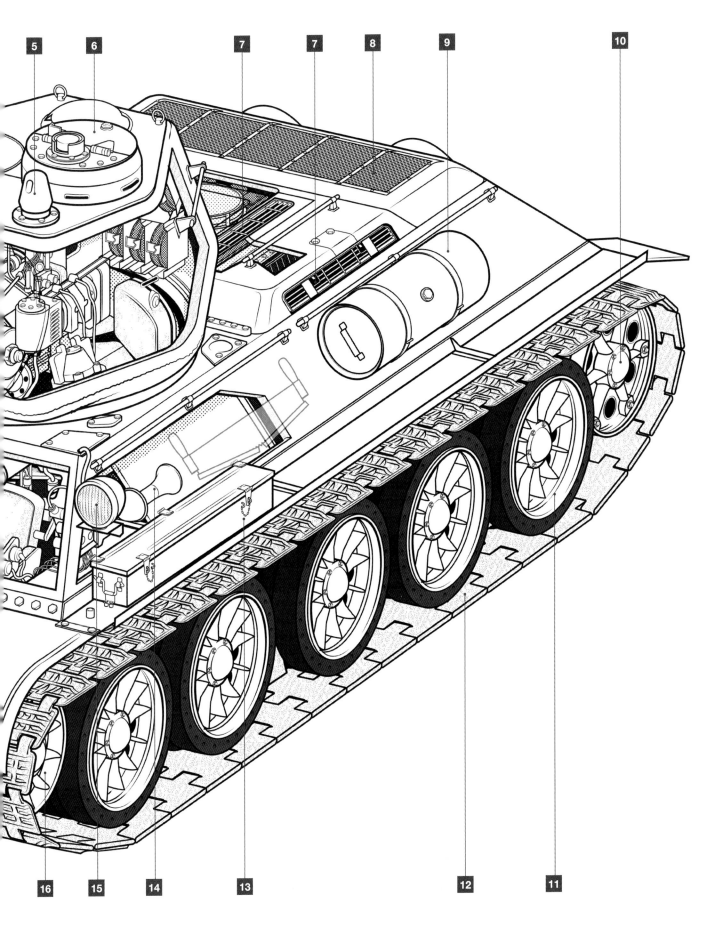

T-34/76 AND T-34/85 ESSENTIAL SPECIFICATIONS

	T-34/76	T-34/85
Crew	4	5
Weight (metric tons)	28	32
Length (hull)	6,070mm	6,100mm
Width	3,000mm	3,000mm
Height	2,604mm	2,700mm
Engine	V-2 diesel	V-2 diesel
Max road speed (km/h)	47km/h	55km/h
Max cross-country range (km)	260km	260km
Main gun	76.2mm F-34	85mm Zis-53 Mod 1944
Secondary armament	2 × 7.62 DGT mg	2 × 7.62 DGT mg
Ammunition	100 × 76.2mm + 1,920 × 7.62	60 × 85mm + 1,920 × 7.62

the permeability to water of the lower hull during water crossings as well as the upper hull during rain. In heavy rain, lots of water flows through chinks/cracks which lead to the disabling of the electrical equipment and even the ammunition.

It was noted by the Russians that in T-34s delivered up until the end of 1941 there was a problem of leakage of water via the driver's hatch as, when closed, it was not sufficiently watertight to prevent the egress of water into the hull.

Hull interior

Within the hull, the space was divided into four compartments. The first two of these housed the crew of four.

■ Moving from front to back the first of these housed the driver-mechanic with the radio operator/machine gunner stationed on his right. Here were to be found the driving controls, gauges and instruments, the engine controls and electrical equipment. Also located here, when fitted, was the radio, which was mounted on the side of the hull to the right of the operator. The driver's hatch was fitted in the glacis immediately to his fore. The door was hinged to the rear on a single multi-lug hinge which extended the full width of the door. The coherence

Indicative of the difficult conditions under which many T-34s were produced in 1942 was the quality of the welds. This was noted in the example evaluated in Great Britain and in the lower hull identified in the analysis of the example of the tank sent to the USA at the end of 1942 for assessment at the Aberdeen Proving Ground. In the report compiled by the evaluation team, it was noted that the main deficiency in the hull was:

BELOW Cutaway plan view of the hull showing the general layout of the engine compartment and running gear.

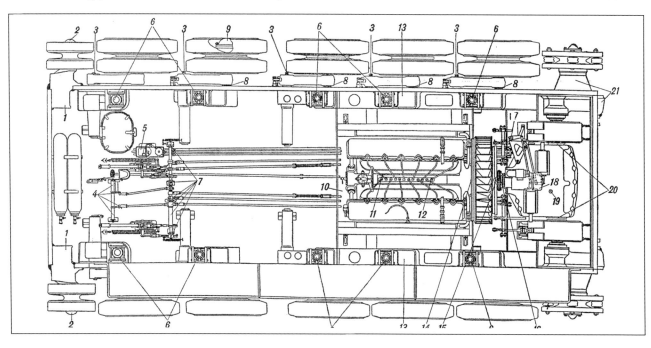

of the thickness of the glacis armour was maintained by armour being welded to the rear of the driver's hatch so that it occupied the space of the hatch when closed down.
- At the feet of the gunner/wireless operator there is an auxiliary escape hatch.
- To the rear of the driver and radio operator/machine gunner came the fighting compartment, wherein was located the tank commander and the loader. Unlike the German medium panzers, the T-34 was never, either in its 76mm or 85mm armed variants, equipped with a turret basket. The two seats in the turret of the T-34/76 and the three seats of the T-34/85 were attached to the turret ring and thus revolved with it.
- Behind the fighting compartment came that of the engine separated from the former by a bulkhead/firewall and to its rear that of the transmission and brake assemblies.

The plan diagram provided opposite illustrates the manner in which the sections within the hull were located and the contents of each.

Turret

The T-34 tank only ever mounted three weapons during its production life – variants of the 76mm and the 85mm cannon and a very small number of the 57mm Zis-2. These were housed, especially in the T-34/76, in a multitude of turret designs. While different in appearance, those on the T-34/76 retained the same two-man configuration with the internal arrangement also being the same. The appearance was dictated by the methods employed by the particular factory producing them so that in the T-34/76, the turrets were either of a cast, welded or stamped construction, whereas those of the T-34/85 were always cast. Rather than describe them all, most are illustrated in the drawings provided of the differing turrets and are also to be seen in the many photographs which are captioned to differentiate between them in the text.

The T-34/76 example supplied to the British for evaluation had a cast turret and the observations made about it are worthy of reproduction:

The turret casting is interesting in view of the high angle of its sides, i.e. 30°, thus conforming with the general high degree of angularity. Although some rolled plates have been welded into the roof, the turret is essentially of the 'all-cast' variety and not of the composite type. The finish of the turret is not of the same standard that British [or US] practice demands – some porosity is evident. This does not necessarily indicate that the casting is inferior from a ballistic point of view.

There is a large hatch provided in the turret roof in the form of a dished door, hinged to the front. The hinges are flush fitting and are welded to the door and bolted to the turret roof. The door is secured in the open/vertical position by a spring loaded catch which engages a lug welded to the turret top. This assembly is mounted externally but may be released from inside of the hinge. The opening is provided with a B.P. hinged flap which is secured by a spring loaded catch operated by a canvas strap.

Pistol ports were not standard on all models of the T-34/76, but were on the T-34/85. Those in the former are conical apertures of approximately 48mm diameter inside the turret and 71mm outside the turret. The ports are closed by conical plugs anchored to the inside of the turret wall by a chain. The plugs are each secured in the ports by a pivoted plate with a slot which engages an annular groove in the plug. The plug is removed by a sharp blow.

(All images British Army, School of Tank Technology except where credited otherwise)

BELOW Turret access hatch.

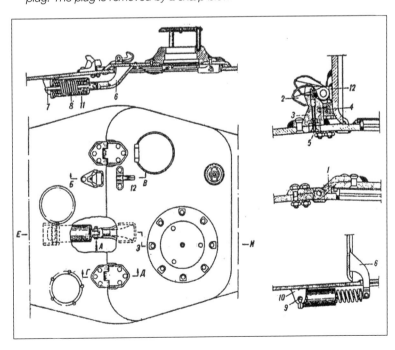

Armour thickness and quality

The effectiveness of the armour on the T-34 hull/turret was not just a consequence of its thickness or the angle at which it was mounted. It also stemmed from the quality of the armour actually employed. On the T-34/76 tested by the Royal Military College of Science School of Tank Technology in late 1943, the hardness of the armour was tested employing the Poldi method and measured on the Brinell scale. The Poldi method was the use of an indenter – this was a small body with a hard tip which functioned as a probe that was propelled against the tank armour. The motion of the probe is the measure of its kinetic energy, which is then converted into a chosen scale of conventional hardness. In this case, that employed was the Brinell scale, where the higher the Brinell number, the greater the hardness of the armour.

It was noted by the School of Tank Technology that the Brinell hardness figures for the T-34/76 were greater than for the armour of contemporary British tanks. This was also the case for the T-34/85 when examples captured in the Korean War were examined in the USA. In general, the armour on Soviet tanks was heat-treated to a very high level of hardness. This had real practical consequences. It was noted by one T-34 battalion commander that as far as Soviet tankmen were concerned, only British tanks had superior armour. He observed that:

If a shell had gone through the turret of a British tank the commander and the gunner could have stayed alive because there were virtually no splinters while in a T-34 the armour would spall a lot and the crew had few chances of survival.

This was explained by the fact that the nickel content in British armour was deemed by the Russians to be very high at 3.0–3.5%. That of the 45mm armour of the T-34 glacis contained only 1.0–1.5%. The Poldi Brinell hardness figures of various armour plates were established as follows:

Glacis plate	354–400
Pannier side plate (nearside)	388–434
Pannier side plate (offside)	387–98
Upper tail plate (outside)	400–10
Upper plate (inside)	389–406
Engine cover plate (cast)	405–7
Turret	370–5
Turret escape hatch (pressing)	390
Gun mantlet	407
The surface hardness of the weld metal by Poldi test is	150–89

Suspension

The suspension follows the original Christie design very closely and rollers and sprockets are employed. Although the British evaluation team did not have a BT to hand to compare the suspension of the T-34 with, it is known that of the medium tank was taken over almost exactly from the earlier light tank, with modifications in the form of stronger springs to allow for the heavier weight. The evaluation team were already familiar with the Christie system as it had been (and was) employed in a series of British cruiser tanks beginning with the Covenanter, Crusader, Centaur running through to the Cromwell and Challenger. Although the Christie suspension had a US provenance, it was not adopted by the US Army, with the Aberdeen Proving Ground report on their T-34/76 noting that it had been 'tested a long time ago by the Americans and

BELOW Leading spring.

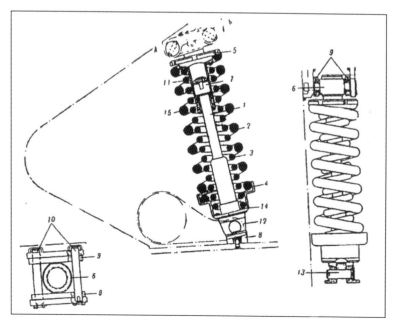

T-34/76 WITH AOHA – ADD-ON HULL ARMOUR

An unusual feature found only on a number of T-34/76 1941/1942 models was add-on hull armour (hereafter AOHA). This was a contingency officially adopted following the inability of industry to execute GKO Resolution No. 1062 of 25 October 1941, which required STZ and Plant 112 (Plant 183 had already been evacuated) to increase the thickness of the hull glacis of the T-34 from 45mm to 60mm. However, there is evidence that Plant No. 183 had started developing and producing AOHA even before the invasion, prompted by the erroneous intelligence that the Germans were fielding more powerful anti-tank guns that could penetrate the 45mm frontal armour of the T-34.

What had prompted this decision was the perception that the Germans were deploying many more PaK 38 50mm anti-tank guns firing APDS (armour-piercing discarding sabot) ammunition, which had the ability to penetrate the glacis of the T-34/76 at a range of 1,000m. This same ammunition was also being utilised by the Panzer III, but in this case the range was reduced to 500m. It was the perception of the vulnerability of the T-34 to these weapons that had driven the GKO resolution. However, owing to the inability of the steel factories to manufacture 60mm armour plate in the quantities required, a fall-back was adopted whereby thinner armour plates were cut to fit and then welded to the front of the 45mm glacis of the T-34. These AOHA plates, varying in thickness from 20 to 35mm, served in practice to increase the frontal armour thickness of those machines fitted with them from between 90mm through to 160mm, taking into account the angle of the glacis. At least seven different styles of AOHA have been identified. It was not a uniform addition to T-34s.

However, these plates showed no uniformity of thickness (as can be inferred from the figures given above) or of design, with every type of AOHA differing in terms of width, length and in the number of segments cut. The AOHA for the tanks manufactured at the STV were produced at the Stalingradsky Sudostroitelniy Zavod (SSZ) – the Stalingrad Shipyard Zavod No. 264 and supplied to the STZ where they were welded on to the glacis on the production

line. Those AOHA used by Plant 112 came from the Kulebaki Works (Zavod No. 178) and Vyksa Works (Zavod No. 177) and were fed into the Gorki tank manufacturing plant. As the first AOHA was designed to fit the T-34 Model 1940 and Model 1941 that came from Plant No. 183, it pre-dates the evacuation of the same to Nizhne Tagil in September/October 1941. The GKO rescinded its order in February 1942 with the two factories fitting them doing so until supplies of the AOHA ran out.

LEFT Even before the German invasion, erroneous intelligence that the Germans were equipping their tanks with more powerful main guns prompted the creation of additional armour plates that could be added to the glacis of the T-34 to 'beef-up' its protection. The AOHA (or Add-On Hull Armour) is seen here on two T-34 Model 1941s built at the STZ and in service on the Kalinin Front in 1942. *(Nik Cornish)*

BELOW In addition to the AOHA to the hull glacis on these two T-34s, there is also AOHA applied to welded turrets. This was an unusual feature mainly to be found on T-34s serving in the Leningrad area. *(RGAKFD Krasnogorsk via Stavka)*

RIGHT The 2nd, 3rd, 4th and 5th springs.

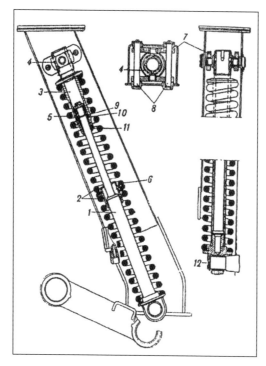

unconditionally rejected'. The T-34 marked the last occasion in which the Christie suspension was used on a Soviet tank, but as we have seen its limitations had been acknowledged by the Russians themselves, when in 1941 a torsion bar suspension was to have replaced the Christie type in the abortive T-34(M). As it was, the designed replacement for the T-34, the T-44 medium tank which emerged in 1944, employed a torsion bar system.

Nonetheless, the Christie suspension of the T-34 operated as follows. Each section should be read in conjunction with its associated diagram(s).

Sprockets

The T-34 has two sprocket wheels at the rear, each of identical design. That being said, at least nine different designs of sprocket wheel have been identified over the production life of the T-34. As with the road wheels, there was no requirement for there to be a set design and, depending upon supply and production conditions, each Zavod often designed and produced their own to accommodate these contingencies.

Each is fitted to the splined end of the driveshaft of the final drive. The wheel is fixed by a ring which is screwed by four studs to the end of the final driveshaft. The securing ring is covered by an armoured cap. The hub of the wheel has a flange. In later models, the rims of the wheels are cast integrally with the discs and the hubs. The discs of the wheel are bolted to the flange. The six large holes in the discs are to permit the egress of mud and snow. To prolong the life of the rims, steel tyres are employed. The shafts are housed between the discs and carry bronze bushes upon which the return rollers rotate. As the sprocket wheels rotate, they drive the tracks by means of the rollers which bear against the guide horns.

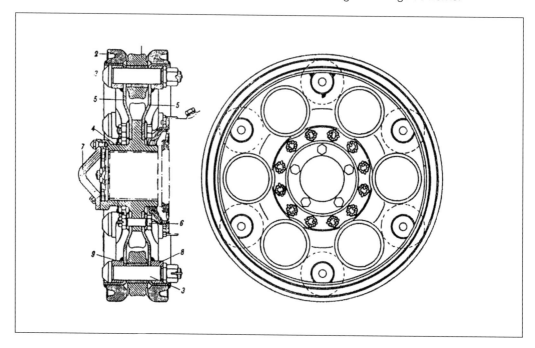

RIGHT Sprocket wheel.

T-34 WALK-AROUND

The T-34/85 that is the subject of this 'walk-around' is to be found in the Tank Museum at Bovington. Unlike the one which was manufactured in Poland, used in the displays in the arena on public days, this particular machine was captured in the Korean War. It is therefore one of the machines supplied by the Soviet Union to the Army of the DPRK (the T-34 in the Korean War is covered on pages 107–110). In addition, a number of the images are taken from the T-34/85 held in the tank collection of the Royal Military College of Science at Shrivenham. These are used to supplement those of the Bovington machine and also to contrast aspects of the design. The T-34/85 at Shrivenham was manufactured at a different factory to the Bovington example.

The walk-around begins with the turret and moves down through the hull, then to the chassis.

ABOVE The T-34/85 on static display in the Bovington Tank Museum in Dorset, England, was a product of the Zavod No. 112 Factory at Gorki and dates from the second half of 1944. Indeed, it is very likely that it was manufactured at the very tail-end of that period as it evidences a number of features to be found on the T-34/85 Model 1945. The NKPA T-34s encountered by the US Army in Korea came from a number of Zavods including Uralvagonzavod No. 183 at Nizhne Tagil and Zavod No. 174 at Omsk. Turrets of T-34/85s were made in a fashion distinctive to the plants that produced them.

(All photographs in this section are Copyright Matt Sampson/Tank Museum except where credited otherwise)

LEFT The turrets of Zavod No. 112-produced T-34/85s were produced from a five-part casting. As can be seen from the image, the welds joining the two parts together are very rough, as is the overall finish of the turret casting itself. The roughness of the part of the mantlet for which the hole for the gunner's telescopic sight can be seen was also a distinguishing feature of T-34/85s produced at this plant, as were the angles of the join for the lower, front and rear parts of the turret casting. The hole in the turret side is where the plug for the pistol port would have been. This was the second version of the T-34/85 turret produced by Plant No. 112.

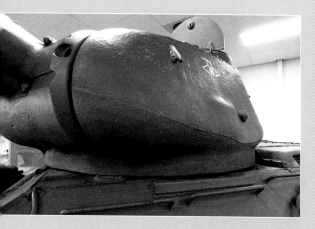

LEFT By contrast, a similar image to the one on the previous page is offered of a T-34/85 produced by the Uralvagonzavod No. 183 plant. One of the distinguishing features of the turret is the slight bulge to be seen before the pistol port. This was to permit the space inside the turret for the fitment of a new electric drive introduced in 1944. What is quite apparent is how much smoother the casting is when compared to that of Plant No. 112. Note the subtle difference in the shape of the lifting hook when compared to those on the Plant No. 112 turret. The projecting flap secured by two nuts on the hull top below the turret is the cover for the diesel inlet, while the flush cover to its fore, secured by three nuts, is an access panel to the road suspension spring. These differed in size and shape, depending on which factory produced the tank. *(Dick Taylor)*

RIGHT The extremely rough casting of the upper and lower turret castings is clear to see. On all T-34/85s produced from mid-1944 onwards, six brackets were welded to the turret rear to enable the crews to attach tarpaulins and other items of personal use. This feature is still to be found on many post-war/modern Russian tanks. The uneven shape of the lower section of the upper turret casting is very apparent, as is the manner in which the workers at the factory employed large quantities of welding metal to mate the upper and lower turret parts together to create an even join.

LEFT On the rear of the turret is mounted the twin domes covering the ventilators. These are now located further back than on the first T-34/85 turrets produced at Plant No. 112. Once again, very heavy welding beads are apparent, but given that they in no way impact on the ability of the turret to function, it really is indicative of the Russian propensity to not see something like this as being of any consequence. There are a number of features on the roof of this T-34 that suggests that while it was produced towards the end of 1944, it mounts a cupola that first came into use on those machines built at the tail-end of 1944 or the beginning of 1945. The turret built by Plant No. 112 in 1944 mounted twin domes on the rear of the turret roof with the cupola normally having split hatches with four hinges, whereas that seen here has a single hatch with two hinges. This was normally found on T-34s where the two ventilators were separated – the so-called 'mushroom turret' – with one remaining on the rear of the roof and the other moved forward slightly offset and placed in front of the loader's hatch. The spaces formerly occupied by the sights have been sealed. Nor does the radio aerial, which would have been placed in the raised circular fitting standing proud of the roof in front of the cupola, remain.

RIGHT The commander's cupola was first introduced on the late models of the T-34/76 and was carried over on to the T-34/85 from the T-43. However, the vision slits did not provide the T-34/85 commander with the degree of vision provided by the sighting devices to be found on the cupolas of German tanks. The principal benefit to the Russian T-34 commander was that he could expose his head and thus acquire a much improved awareness of the situation around his vehicle. While the body of the cupola below the hatch was fixed, the hatch itself was traversable.

ABOVE This cover was designed to prevent the ingress of shrapnel and rainwater into the gap between the mantlet and the body of the turret. It moved up and down when the gun was elevated and depressed.

ABOVE In this view of the whole of the glacis plate, 47mm of armour angled at 60 degrees, the angled mudguards over the front of tracks can clearly be seen. These are a further feature of the Plant No. 112 production for the second part of 1944, although not unique as Plant Nos 174 and 183 had also adopted this style. Prior to that date the mudguards were curved. The four bolts on the lower glacis are for the carriage of spare track links. The weld join of the lower part of the glacis to the angled hull nosepiece is apparent.

ABOVE The hatch in the glacis plate was the driver's primary point into and out of the hull. One of the observations of the British evaluation was that while this was satisfactory from an exit and mechanical point of view, it did detract from the efficacy of the ballistic standard of the glacis plate layout. Above the casting marks on the driver's hatch are to be seen the two armoured covers that protected the driver's prism sights which were introduced in early 1942 to replace the driver's periscopes.

ABOVE Here the driver's hatch is open to its full extension. It is hinged to the rear with a multi-lug hinge. It has two locks, which can clearly be seen, that can be opened from within by pulling a strap connecting both locks. To assist opening the hatch, a balancing mechanism is provided, together with a catch and hand wheel for supporting it in the open position. The image clearly shows that the thickness of the armour of the hatch was greater than that of the glacis plate. *(Dick Taylor)*

ABOVE The armoured hood for the hull machine gun was welded over the space cut into the glacis to accommodate it. The Degtyarev 7.62mm light machine gun itself was placed on a ball mount which is protected by an armoured cover, through which the barrel of the weapon projects. The hole above the barrel is for the open sight of the machine gun.

ABOVE This view across the engine deck of the Shrivenham T-34/85 permits a detailed view of a number of features. On the right is the rear of the turret and what is noteworthy here is the much smoother cast finish. Above the air louvres and on either side of the engine deck are two covers for the refuelling points of the rear fuel tanks. The turret would need to be turned to 90 degrees to permit the hatch on the top of the deck to be opened. It was this hatch that gave access to the V-2 diesel engine. The two large 'handles' on the deck are for infantry being carried by the tank in the 'desant' role. *(Dick Taylor)*

RIGHT A reduction of the T-34/85 internal fuel load from 610 to 545 litres saw a corresponding rise in the amount carried externally in three cylindrical fuel tanks. Collectively, these three tanks carried 270 litres of diesel fuel. They were constructed of sheet metal.

LEFT The hinge type on the Shrivenham T-34/85 places it as being produced by Plant No. 112, whereas the turret is a product of Plant No. 183. The two covered electrical leads emerging from above the round hatch would connect to two small drums which were the Bsh smoke generators that would be attached to the lugs on the top edge of the rear plate. Note that this T-34 does not carry the attachment points for the box for the crew's effects, as seen on the Bovington example. To gain access to the interior of the rear hull via the round hatch would require the seven large nuts to be undone.

FAR LEFT Opening the rear round hatch allowed access to the T-34's transmission section. What can be seen here is the electric starter, mounted on top of the gearbox.

LEFT Unlike the differential covers on some T-34s, the two on this machine are smooth castings. The welding on this part was well executed.

RIGHT The road wheels on the Bovington T-34 are characteristic of those fitted by Plant No. 112 for the period 1944–45. There are five retaining nuts on the central hub, six on the inner part of the road wheel and ten on the outer ring. The drive sprocket is not a common type – the latter in almost all cases having either five or six holes and not the ten seen on this example. In profile can be seen the two-part tank track. During its production life the T-34 employed many different track types; at least 16 have been identified. The type used here is a variant of the cast 500mm waffle track of 1944. It is in two parts with the large tooth of a distinctive shape carried by every second link which also houses the two track pins. This track was used on late and post-war T-34/85s and the SU-100 tank destroyer.

LEFT Included by way of contrast to the previous image are these road wheels that were fitted to Plant No. 112 T-34/85s from the 1944–45 period. This is described as being the 'full spider wheel'. Unlike those on the Bovington T-34, these road wheels feature 12 holes designed to assist the ejection of mud that accumulated between the inner and outer wheels. The tracks are of the same waffle track type seen on the Bovington machine.
(Dick Taylor)

RIGHT The tracks on the T-34 were made from cast manganese steel. The multiplicity of types fitted to the T-34 over its production life is once more a reflection of the many plants that produced it as well as being an evolutionary process – new track types were developed to cope with combat conditions. The track pins on the T-34 were not retained by any fitting – they were left unsecured. The purpose of the projecting shaped section welded to the hull and stationed above the track was to push back the track pins automatically when they started to work their way loose. As the track moved around, any projecting pin would come into contact with the shaped projection and be pushed back in. A rather simple, but elegant solution to an ever-recurring problem.

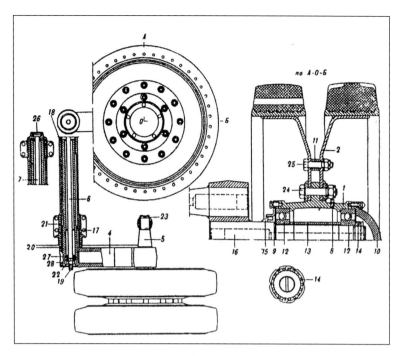

ABOVE Road wheel.

Road (bogie) wheels and cranks

All road wheels fitted to the T-34 (of whatever model) were of the same size. However, both the 76 and 85 variants utilised a multitude of wheel designs and many of these found their way on to T-34/76s, in particular a 'mix and match' configuration, with no fewer than 23 different designs of road wheel having been identified as having been used over the production lifetime of the tank. Two of those types are illustrated in the photographs of the walk-around of the T-34/85 in this book. Such differences were also found on the self-propelled guns and tank destroyers. Where possible, attention will be drawn to these differences in the photographs in this book.

In the hub (1) of the road wheel are pressed in two ball bearings (12). Between them there is a distance piece (13). The road wheel is attached to and fixed by a nut (14) with a lock. The hub of the road wheel is covered by an armoured cap (10). To the internal end of the hub is fastened a labyrinth seal consisting of two rings (9) and (15). On the outside of the hub there is a flange. The plates are fastened to the flange by bolts (24). Between them are plates supported by bolts (25) which rest in bushes (11). On the plates of the road wheel are pressed and welded steel rims with rubber tyres.

The cranks serve to connect the road wheels to the hull of the tank and to the springs. Each crank consists of a suspension arm and two axles (ie the suspension arm axle and the axle of the road wheel). The axle of the suspension arm rests in the hull in cast iron or bronze bushes. The road wheel is carried on its axle by two ball bearings. When movement of the road wheel takes place as it goes over an obstacle, the crank rotates around its axle in the hull and compresses the spring. On each side of the tank there are five cranks – one for each road

RIGHT This thoroughly demolished T-34/76 – the remains of a hard-edge *Gaika* turret is to be seen in the left foreground – is useful for allowing us to see how the vertical columns within which the springs for the individual road wheels were accommodated within the hull of the tank. These intruded into the crew space with the cells for the diesel fuel being located between them. *(Author)*

wheel. The cranks of the road wheels, with the exception of the front ones, are of exactly the same design. The axle of the road wheels, the axle of the suspension arm and the trunnion for the attachment of the springs are pressed into the suspension arm and welded.

The opposite cranks on both sides of the tank are linked by bolts which safeguard them from lateral displacement. The cranks of the front bogie wheels are made in one piece with the axles, which are carried in the front of the hull.

One particular design of road wheel is worthy of note. Towards the end of 1941, a serious shortage of rubber was experienced, such that rubber-tyred road wheels of the type described above were replaced by an all-steel type. Designed at the STZ in October 1941, production of this type soon spread to all factories producing the T-34. The appearance of this type of road wheel was very distinctive and its sound was even more so! Even when rubber supplies improved and it was possible to once more produce rubber-tyred road wheels, it was common, especially in 1942 and early 1943, to see T-34/76s with a mixture of both types. Rubber-tyred road wheels would be employed on wheel stations one and five with the three intervening stations utilising the steel type.

Engine

Originally designed by A. Morozov and a team at Zavod in Kharkov in 1938, the 500bhp diesel engine has proven to be one of the longest serving of all engine types used for powering tanks. Indeed, the latest iteration of the engine used originally to power the T-34 is still being employed, albeit in a much-modified and more powerful form, on the contemporary T-90 main battle tank.

Basic specification

- Type: V-2. 60 degree V-12 cylinder, water-cooled CI
- Bore: 150mm
- Stroke: 180mm (articulated rods), 186.7mm (master rods)
- Capacity: 38.8 litres
- Compression ratio: 15:1 (articulated rods), 15.8:1 (master rods)
- Rated power: 500bhp at 1,800rpm.

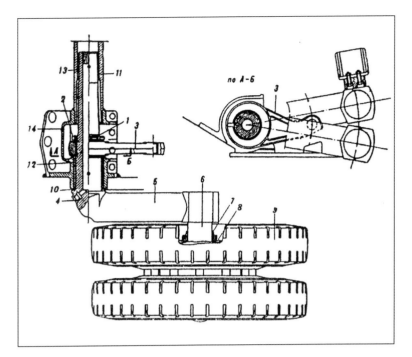

The engine is situated in a separate compartment at the rear of the hull with a radiator and a fuel tank on each side. It is of a very light construction, with the cylinder heads, blocks and the crankcase being of aluminium alloy. The blocks are detachable and are fitted with wet liners. The combustion chamber is of an open type with the injector nozzle centrally in the head. To gain access to the injectors it is necessary to remove the valve covers. Four vertical valves per cylinder are

ABOVE Leading road wheel.

BELOW The 500hp V-2 diesel engine of the T-34 series as seen when the upper deck engine hatch is opened. *(Nik Cornish)*

ABOVE The 500hp diesel engine of the T-34.

BELOW V-2 engine cross-section.

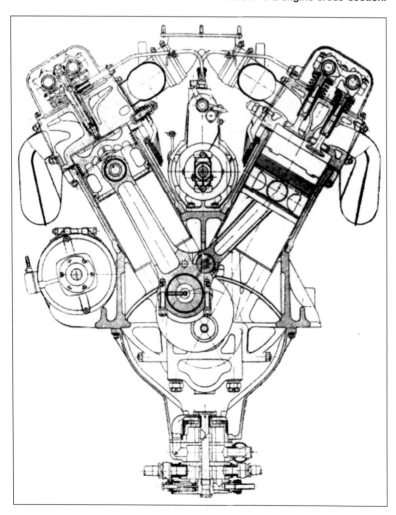

operated directly by the camshafts. The pistons are of hot pressed aluminium alloy and the connecting rods are articulated of 'H' section and machined all over. Copper lead bearings are used in the big and main ends, and a ball thrust race is located on the driving end of the crankshaft. The crankshaft is supported in eight bearings and is fully machined. No torsional vibration damper is fitted. The crankcase is split along the crankshaft centreline, but extends below this level to support side thrust on the main bearings. A bevel on the free end of the crankshaft drives up to the camshafts, injector pump and governor, air distribution valve and dynamo. Driven also from this bevel and located on the sump are the water pump, oil pump and fuel feed pump. The engine is mounted rigidly at four points on longitudinal bearers welded to the hull floor. The mounting is not particularly rigid and is not braced to the hull side plates.

It was observed by the evaluating team that:

The quality of the workmanship varies considerably. Whilst the highly stressed parts have a finish comparable with British aero engines of a moderate output, the sand castings by contrast are exceptionally rough. In spite of this, however, the latter appear to be sound, there being no signs of porosity or blow holes on the machined surfaces. Most of the important bolts and studs are stress relieved and ground, and on a few components the standard of finish is very high. The large number of inspection stamps on certain components is particularly noticeable.

Air cleaner

An oil bath-type air cleaner is mounted centrally above the engine. It has a very shallow element of loosely packed steel wire. The standard of workmanship in this component is particularly crude. The cleaner must be removed to obtain access to the fuel injection pump.

The British perception of the very poor quality of the air cleaners was also picked up in the Aberdeen Proving Ground report on the T-34 they received for evaluation. Indeed, it was the inadequacy of the air cleaners – they were extremely rudimentary – that made a significant contribution to the limitations of the T-34's diesel engine. It is for this reason that observations about the problems with the reliability of the V-2 engine have been subsumed under this section dealing with the air cleaners.

Indeed, the limitation of the T-34's diesel engine – for example the one tested at the Aberdeen Proving Ground seized after 72.5 hours, even though much care had been spent on its production and preparation before being sent to the USA, and was maintained while there by a Soviet mechanic sent for the purpose – was far greater than was the norm for T-34 tanks serving in Russia. Average engine life for the T-34 in 1941 was as low as 100 hours. By 1944, this had increased from between 180 and 200 hours, still well below the sort of figures assumed of the US M4, even with its many engine types. Indeed, according to N. Fedorenko, the Head of the Armoured Directorate of the Red Army, the average mileage of the T-34 to overhaul throughout the conflict did not exceed 200km. Given the many tens of thousands of T-34s produced between 1941 and 1945, this figure serves to indicate just how many must have become unserviceable at very low mileages. Nonetheless, Fedorenko also noted that this figure was considered sufficient by industry and the Red Army.

Continuing with their observations about the air cleaners, the US team noted that 'Wholly inadequate engine intake cleaners could be expected to allow early engine failure due to dust intake and resulting abrasive wear.' It was further noted that:

> [T]he air cleaner doesn't clean all of the air which is drawn into the motor. Its capacity does not allow for the flow of the necessary quantity of air, even when the motor is idling. As result, the motor does not achieve its full capacity. Dirt getting into the cylinders leads them to quickly wear out, compression drops and the engine loses even more power.

Indeed, this became a self-fulfilling prophecy with the Aberdeen Proving Ground's example, in that its engine gave out after being driven for 343km. It could not be fixed with the cause being traced to the said ineffectual air cleaners.

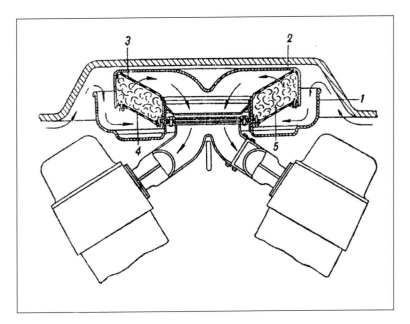

ABOVE Air cleaner.

A large quantity of dirt had made its way into the engine and a breakdown had occurred, in consequence of which the pistons and cylinders were damaged to such a degree that they were impossible to fix.

This then would seem to add credence to Fedorenko's observation about the low mileage that could be driven before overhaul. It would also account for the sheer necessity to produce so many T-34s, as in practice the service life of each example, if not dictated by being destroyed on the battlefield, was certainly limited in time and range by the extremely short distance it could drive before the engine seized and became irreparable.

A Russian analysis of this US assessment noted in language more appropriate to the Soviet mindset wherein any problem was attributable to sinister human activity rather than to technical limitation, that: 'The deficiency of our diesels is the criminally poor air cleaners on the T-34. The Americans consider that only a saboteur could have constructed such a device.'

One wonders, if following receipt of this document, the NKVD went to the places where these air filters were assembled in search of these 'saboteurs'.

That being said, new 'Tsiklon' (Cyclone) air filters were fitted as standard with the production of the T-34 Model 1943. However, the US perception was that the situation with the transmission had not improved substantially when another US team had the opportunity to examine a T-34/85 built in 1945, and captured later in Korea. The evaluating team noted that:

Wholly inadequate engine intake cleaners could be expected to allow early engine failure due to dust intake and resulting abrasive wear. Several hundred miles would probably be accompanied by severe engine power loss.

It should be stated, however, that the US team was applying the standards that would be expected of US tank engines – the various types being used in the differing models of the M4 series at the time averaged hundreds of hours. As we have seen, for the Russians, having a T-34 engine that could last 'several hundred miles', was something that they would have found to be quite remarkable!

Injector equipment

Bosch-type injector equipment is used. The injection pump is a 12-cylinder inline with 10mm-diameter plungers; in design it is basically similar to the C.A.V. B.P.E. series pumps with a centrifugal governor incorporated at the rear. It is situated centrally between the cylinder banks and driven through a Bosch-type fibre coupling.

Exhaust

The exhaust gases from each bank are conveyed via the spate manifolds, which are pressings from sheet steel. They are manufactured in two halves and welded together.

Systems

Cooling

The centrifugal water pump is situated on the engine sump and is driven at 1.5× engine speed. Water is drawn from the radiator and delivered to a single inlet at the base of each cylinder block. From the block it passes into the head through a number of small holes below the valves, and is returned to the radiator via a single port in each head. A remote greaser for the pump is mounted in the fighting compartment on the engine bulkhead and is operated by a turnbuckle.

A gilled tube radiator is mounted at each side of the engine compartment – they are inclined towards the top of the engine at an angle of 30 degrees. The header tanks are coupled by a pipe in the centre of which is the filler cap incorporating a pressure relief valve. Steam is taken off the cylinder heads via small-diameter relief pipes, one for each bank, which couple the return pipes to the filler 'T' piece.

A radial flow fan is mounted on the flywheel and circulates air drawn through the louvres at either side of the engine through the radiator matrices and round the engine. The air is exhausted through external louvres at the rear of the superstructure, having first circulated round the gearbox and steering units. The outlet louvres may be completely closed, or the degree of opening adjusted, by a lever to the left rear of the driver.

Lubrication

The triple gear-type oil pump is mounted next to the water pump and driven at 1.725× engine speed. Two scavenge pumps draw it from each end of the sump and deliver it to two oil tanks. Each tank has a capacity of approximately 50 litres and is situated on either side of the engine between the radiators and hull side plates. There is no oil radiator in the system. The pressure pump delivers oil into the free end of the crankshaft after it has passed

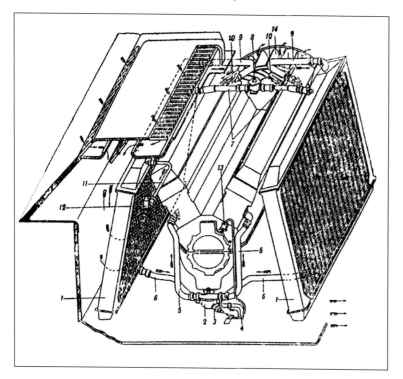

BELOW Engine cooling system.

through a clearance-type filter incorporating a bypass valve. The filter is very inaccessible and removal of the elements presents considerable difficulties. The Aberdeen Proving Ground's General Impressions noted that the oil found in the system as received corresponded roughly to an SAE 50 lube oil.

Fuel

Eight fuel tanks are provided having a total capacity of 610 litres (135 gallons). The respective capacity of the tanks is as follows:

- One wedge-shaped tank above a rectangular tank on each side of the fighting compartment. Fuel carried = 150 litres each tank for total of 300 litres.
- One tank at each side of the rear compartment over the final drive units.
- Fuel carried = 80 litres each tank for a total of 160 litres.
- One tank at each side of engine compartment forward of lubricating oil tank.
- Fuel carried = 75 litres each tank for a total of 150 litres.

A filler cap is provided to each upper forward tank, the lower tanks being fed by gravity from the upper ones. The central tanks in the fighting compartment are coupled to the forward tanks and have no independent filler caps. The filler caps for the forward tanks are accessible from the top of the hull on removal of the BP covers, each secured by two hexagon head set screws. A common filler, situated at the top of the

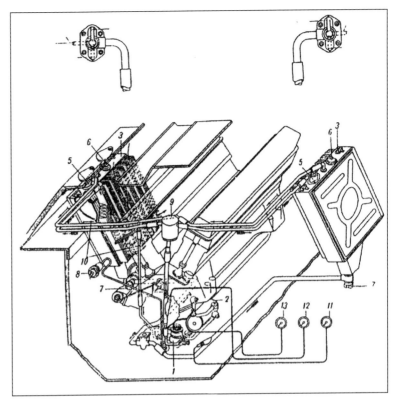

ABOVE Lubrication system.

LEFT This T-34 has been destroyed by a flank shot from a German gun penetrating the fuel tanks on the nearside of the crew compartment. The rupturing of the side armour permits a view of the rear fuel tanks carried on either side of the engine compartment at the rear. The tank itself has AOHA that had been welded to the turret and was a feature of some T-34s employed in the Leningrad area in 1941–42. *(Nik Cornish)*

offside tank is provided for the rear tanks (which are coupled). To reach the filler it is necessary first to open the hinged cover over the air louvres (see photo top right on page 128 in the walk-around) at the rear of the vehicle and then to remove a BP cover, similar to those for the forward tanks. Removable gauze filters are fitted at each filler. A ring halfway up the filters marks the maximum level to which the tanks may be filled. A composite dipstick is provided in the vehicle equipment to measure the quantity of fuel in the tanks. An air distributor cock, mounted on the floor on the left of the driver is provided, and is turned to one of three positions to ensure that the fuel lines are air-free. Pressure is introduced by means of a hand pump situated by the driver's left foot controls.

In the T-34/85, the internal fuel load was reduced from 610 litres carried by the T-34/76 to 545 litres. However, with the employment of up to three external cylindrical fuel tanks, a further 270 litres could be carried. These tanks were also used by the T-34/76 Model 1943. Prior to that model, when extra fuel was carried externally, it was in two types of box tanks mounted on the rear plate. The type of fuel employed by the T-34 is described as DT diesel fuel or E-type gas-oil.

Electrical

The electrical equipment operates at two voltages: 24V for the starter motor and turret traverse motor and 12V for the lighting and communication equipment. The wiring is single-pole earth with negative earth return, a single-pole earth isolating switch being incorporated in the battery earthing lead. The switch is mounted on the hull wall to the right of the wireless operator.

All cables are screened with a braided wire sheath earthed by clamps at the end of each cable run, and except for a few short lengths all cables are carried in pressed steel ducts bolted to the hull.

The 12V services are taken from one bank of accumulators only, when the earthing switch is closed. One internal lamp and an inspection lamp socket have double-pole wiring across

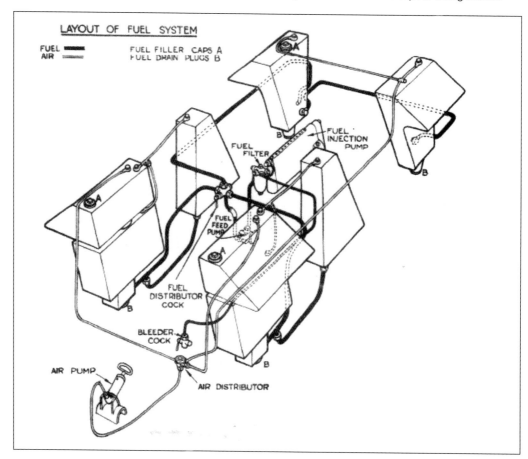

RIGHT Fuel system schematic.

the terminals of this accumulator, and these components are unaffected by the position of the earthing switch.

Accumulators

Four 12V lead acid batteries of about 120amp/hr capacity are mounted in the engine compartment in two banks on either side of the engine. The accumulators are accessible through doors in the bulkhead, and their removal is extremely laborious, much of the fighting compartment stowage having to be removed before the accumulators can be pulled into the fighting compartment for removal through the turret hatch.

Rotary base junction

The feed to the turret services are taken through a fuse box in the left-hand corner of the fighting compartment, containing three cartridge fuses and one wire fuse, then through a rotary base mounted in the middle of the fighting compartment floor. The case of the base consists of an iron casting, the fixed portion being held to the turret floor by four bolts.

The current from the battery is fed and returned from the turret components by three large slip rings – the top and centre carry respectively +24V and +12V. The lower ring is the earth to the turret and hull. Six small rings carry the intercommunication circuits.

Transmission

Although there were a number of problems with the T-34, its Achilles heel was undoubtedly the transmission. In the opening months of the conflict it was so unreliable that it was not uncommon for the Germans to find early model T-34s abandoned because of mechanical failure with a spare transmission strapped to the engine deck. The T-34/76 evaluated by the British in 1943 employed the standard transmission as used on the 1942 model, which differed little from that of the earliest variants.

Clutch

A multi-plate dry clutch is mounted on splines on the end of the crankshaft. The fan and

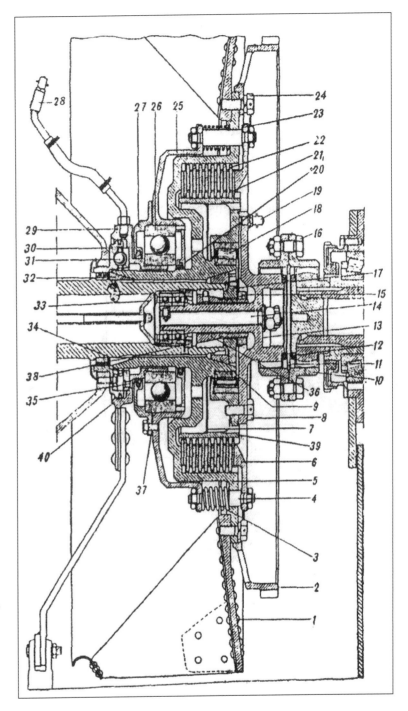

ABOVE Engine clutch.

starter ring gear are bolted to the clutch driving member. Eleven driving plates are located in internal splines in the driving member and ten driven plates in external splines on the driven member. The driven member is supported on the driving member by the large single-row ball race. A similar ball race carries a cage situated between the clutch plates and the engine to release the spring pressure on the pressure plate and disengage the clutch. The withdrawal

RIGHT **Mindful of how problematic the transmissions were in the early T-34 Model 1940s and how easily they ceased to function after being driven even a short distance, Russian tank crews strapped a spare to the engine deck. Clearly on this occasion, assuming this T-34 broke down, the crew did not have the time to employ it and was captured in this state by the Germans after it had been abandoned.**

mechanism itself consists of a type of ball-thrust race situated between the withdrawal cage and the engine. One member of this race is bolted to the crankcase and the other may be rotated by the clutch withdrawal lever. The balls, however, do not run in an annular groove, but each in a separate depression. Thus, when one member is rotated by the withdrawal lever, the balls ride up these depressions and force the withdrawal cage outwards. Mechanical linkage connects the withdrawal lever with the clutch pedal.

Gearbox

The drive from the clutch to the gearbox is taken through a splined muff coupling, the splines on the gearbox input being spherical to allow for a certain amount of misalignment.

The gearbox itself is a simple four forward and one reverse speed sliding mesh type, operated through a linkage from a change-speed lever to the right of the driver. The primary shaft is driven at right angles from the input shaft through a pair of spiral bevels. The input shaft is supported on a parallel roller race, a ball-thrust race and a taper roller race. Both primary and secondary shafts are supported, at their centre, on double-taper nose races and at each end of parallel roller races. First and second gear driving pinions are integral and slide on the secondary shaft. Third and fourth gear driven pinions are integral and slide on the secondary shaft. Reverse pinions engage with the first-gear pinions. The casing is of cast aluminium split along the centreline of the shafts. A filler cap is provided in the top of the casting. No dipstick is provided, but the service handbook specifies a minimum level of 40mm. No oil pump is fitted, the lubrication being entirely by splash.

A somewhat unusual locking device to prevent the gears jumping out of engagement is

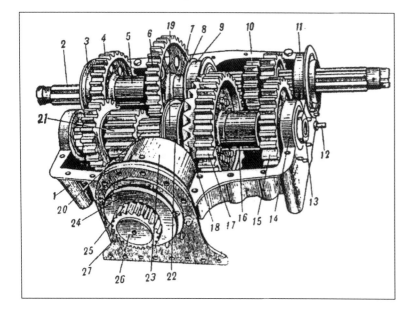

BELOW **General view of the gearbox.**

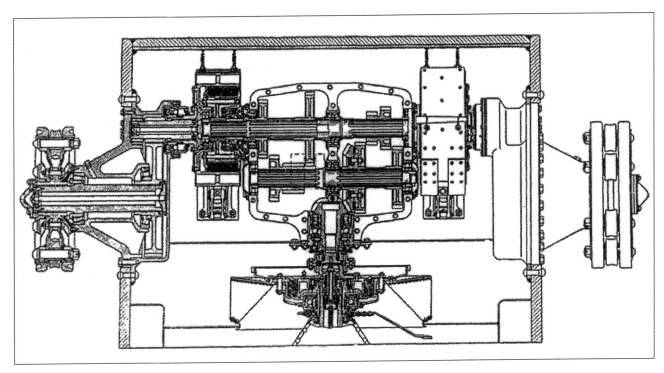

ABOVE Transmission.

incorporated in the change-speed mechanism. It consists of spring-loaded plungers actuated by balls engaging dimples on the selector control rods, and a rotating circular locking plate, with holes appropriate to each plunger. A trigger on the change-speed lever rotates the circular locking plate, thus bringing the plungers into alignment with the holes. Upon engagement of the selected gear, the appropriate plunger enters its respective hole and the gear and controls are positively locked. It will be seen that movement of the change-speed lever for the selection of any forward speed necessitates depression of the trigger. To engage reverse gear, the trigger is not operated until the gear is engaged, when it is depressed, and retained in this position by a hinged catch on the change-speed lever.

Alongside the problematic nature of the air filters, the Aberdeen Proving Ground also noted the issues that arose with the transmission.

. . . on the T-34 the transmission is also very poor. When it was being operated, the cogs completely fell to pieces (on all the cogwheels). A chemical analysis of the cogs on the cogwheels showed that their thermal treatment is very poor and does not in any way meet American standards for such mechanisms.

Nor had the transmission improved by any significant degree in the T-34/85 examined some years later after it had been captured in Korea.

. . . the transmission by American standards had already failed, although with extreme care it could have been used further. Teeth ends on all gears were battered as a result of clash shifting. Many pieces of gear teeth had been broken off and were in the transmission oil. The failure is due to inadequate design, since excellent steel was used throughout the transmission.

BELOW Transmission and final drive.
(Nik Cornish)

The proviso in the previous observation is of course 'by American standards' and herein lies the essential point. The T-34 was not, nor could have been, designed to American standards. It was the product of a lower level of technology, notwithstanding the outstanding merits of its basic design.

Steering

The gearbox secondary shaft is extended each side to carry the driving members of the steering clutches. Each clutch consists of 15 driving and 15 driven plates, the design being basically similar to that of the main engine clutch with a similar type of withdrawal mechanism operated mechanically through toggles and rods by the steering levers. When the brakes are free, the bands are held off the drums by four small tension springs. An interconnected foot pedal operates the same bands on the steering brakes through a transverse compensating shaft. A hand-operated retaining catch engaging with the footbrake pedal provides for parking.

Final drive

The final drive unit consists of a single straight spur reduction gear mounted in the armoured housings at each side of the tail of the vehicle. Lubricant is introduced to the units through filler plugs located near the top of the housings, outside the hull side plates.

Idlers

The twin idler wheel is of one-piece construction, similar in design to sprockets but without the rollers. Tyres are not fitted.

Tracks

Each track consists of 74 manganese steel track links. Every other track has a guide horn which serves to mesh with the sprocket wheel and hold the track in position. The track links are hinged to one another by 148 pins. The track pins have round heads at one end and are 'hot-cropped' at the open end. The pins themselves have no retention device and are retained by the fitting of bevelled wiper plates welded on to the hull sides at the rear of the vehicle.

It was noted by the British evaluation team that on receipt of the vehicle several pins on each side had slipped out beyond the point at which the wipers were effective. This condition may have been caused by vibration set up during transportation and may not occur when the tracks are under load.

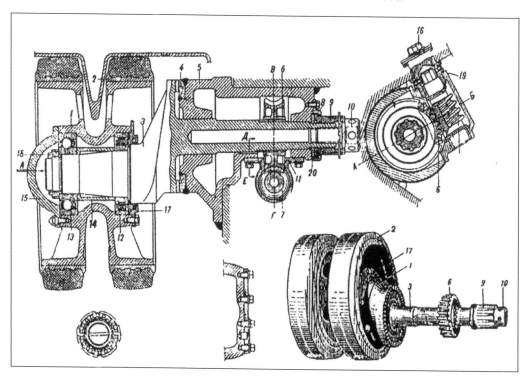

RIGHT **Idler wheel and track adjuster.**

Track adjustment

Adjustment of the tracks is effected by means of the cranked mounting of the front idler wheels. The shaft of the idler crank is housed in a bracket welded to the hull. A concentric toothed ring is formed on the web of the crank. The teeth on this ring engage teeth on the mounting and the adjustment is made by the loosening of a splined nut which disengages the teeth, and rotation of the shaft by means of a worm and a worm-wheel. The worm-wheel is splined to the shaft and the worm rotates in a mounting on the glacis plate. A square boss protruding through this plate at each side is engaged by the adjusting tool. A lock bolt is provided to ensure the security of the splined nut when the adjustment is complete.

Towing and lifting eyes

Combined towing and lifting hooks are welded to the glacis plate at each side. These incorporate a spring-loaded safety catch to prevent the towing cable from jumping the hook.

Heavy towing eyes are welded to the hull. Two towing cables and shackles are carried on the nearside track guard. Lifting eyes are fitted at three points on the turret.

Wireless and intercom

The tank is equipped with wireless and intercom, the latter system being independent of the wireless and capable of operation with the wireless removed (or, as was the case earlier in the war, not fitted).

Wireless

The wireless installation consists of a receiver and sender mounted in a pannier to the right of the front gunner together with a control panel which carries the rotary transformer switch. Two sockets are provided in the face of the control box either for direct connection of the headsets and microphone or for the connection to the operator's intercommunication box. The receiver has only one tuning knob and one volume control. Two flick frequencies are obtainable by adjusting points on the knob. The sender has one tuning knob and a variometer working in conjunction with a tuning lamp.

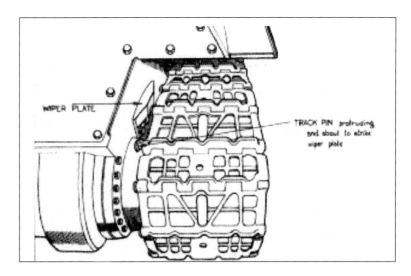

ABOVE Track pins.

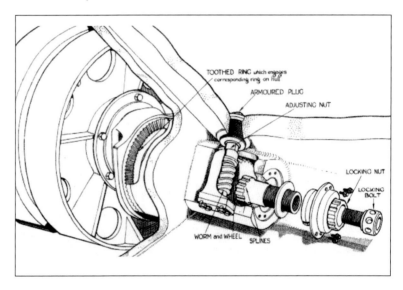

ABOVE Track adjustment.

BELOW Combined towing and lifting hooks.

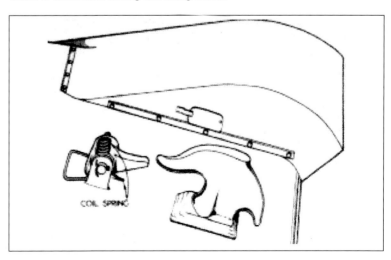

RIGHT Wireless installation.

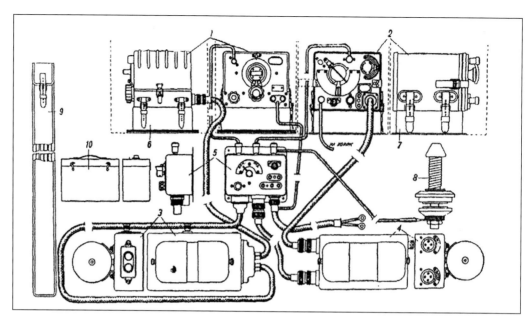

BELOW Seen on the right from the gunner's seat is the radio, as described in the test. In practice the range and reliability was never what it was supposed to be. Note also the ball mount in the glacis for the Degtyarev light machine gun to be placed (see page 153). The box attached to the upper glacis is the intercom. Directly beneath it with the balled palm grip is the gear shift, which, while primarily the driver's task to change, could often be so difficult to change that it required both the driver and the gunner to work together. The pillar housing in the corner of the hull contains the coiled suspension spring. *(Nik Cornish)*

The two rotary transformers are mounted below the sets and are connected to them and to the control box by braided metal-sheathed cables terminating in plugs for engagement with sockets in the various components. The aerial is mounted on an isolated base on the right-hand side plate at the front and can be dipped by a lever behind the wireless sets inside the tank.

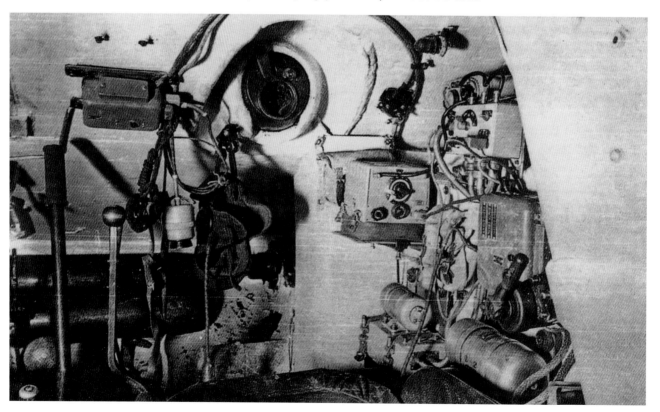

The intercom box is to the operator's right and can be connected to the wireless control box by plugs, in which case the operator's headset and microphone are plugged into the intercom box. The box carried a three-way switch, a push-switch and a lamp. The three-way switch has positions 'Set to Self', 'Intercom' and 'Sets to No. 3', enabling the operator to put the commander on the air. Apart from the operator only the commander and driver are provided with headsets and microphones. The driver's intercom box is equipped with a calling push only and he can speak and hear on the intercom.

The commander's box has a two-way switch, a push-switch and an indicator lamp, the two positions of this switch being either 'Wireless' or 'Intercom'.

The headsets consist of two receivers arranged to be carried in the pockets of the crash helmet. The microphone is a high-resistance type with a wooden mouthpiece and a 'press to speak' switch.

The experience with the intercom system was hardly as positive as explanation suggests. In practice to surmount its unreliability, particularly in the early years of the war, the tank commander and driver had to come up with an agreed method that could communicate the former's intention without the use of the intercom. One means whereby this was carried out was with the commander resting his boot-shod feet on the shoulders of the driver and 'he would press on my left or right shoulder and I would turn to the left or right accordingly'. In the same fashion, albeit by using his hands, the commander could communicate with his loader by an agreed series of gestures. 'If I poked my fist under the loader's nose he knew he had to load an AP one. If with an open palm, with fingers apart, a fragmentation one.' (Quotes from Drabkin and Sheremet, *T-34 in Action*.)

BELOW In the Aberdeen Proving Ground's evaluation report on the T-34 it was noted that the driver's position was not comfortable to sit in. In front of him are the two steering controls and foot pedals for the clutch (on the left) and brakes (on the right). Just to the right of the latter is the driver's foot throttle pedal. The box with wires coming out of it to the upper left of the driver's seat is the electrical distributor panel. A little beyond that but out of sight from this perspective is the starter button. In front of the driver and mounted at eye-level just beneath the driver's hatch is the instrument panel. Mounted on the glacis are two air bottles to aid in engine start-ups. The large black cylinder on the top left is the counterweight assembly that was employed when the driver had his hatch open. *(Nik Cornish)*

Chapter Six

T-34 weaponry and firepower

Although the 76mm-armed T-34 was formidable in 1941, the appearance of more heavily armed German tank designs in mid-1943 prompted the development of a new turret equipped with a longer 85mm gun. This kept the T-34 effective to the end of the war in Europe and for many years after.

OPPOSITE It was on Stalin's insistence that the T-34 armed with the 76mm F-34 weapon be maintained in production to maximise output, even after it had become apparent to commanders in the field that new German weapons and anti-tank guns of greater calibre were leading to high losses. Only after the Battle of Kursk which, although a Soviet victory, resulted in extremely high losses of T-34s, did he relent and permit the tank to be equipped with an 85mm weapon in a larger, three-man turret. Here a column of T-34/85s of the 1st Mechanised Corps lines the Kaiser-Wilhelm Strasse in Berlin on 25 April 1945. *(RGAKFD Krasnogorsk via Stavka)*

During its production life, be it in the Soviet Union or when built under licence in Poland and Czechoslovakia, the T-34 tank only ever mounted three calibres of weapon. The calibre of weapon carried indicated the type of T-34.

- The first and most numerous of the type was the T-34/76 where the designation 76 indicated that it carried either a 76.2mm L-11 or 76.2mm F-34 gun as its main armament.
- The next in chronological order was the T-34 mounting the 57mm gun derived from the Zis-4 anti-tank gun. Only a very small number were built.
- The first T-34/85 was designated the Model 1943 and mounted an 85mm D-5S gun. This weapon was then superseded by the superior Zis-S-53 85mm weapon from March 1944. This was to remain the primary armament of the T-34 until such time as it ceased production.
- All models of T-34 built carried two Degtyarev DT 7.62mm machine guns. One in the hull and the other co-axial with the main gun.

T-34/76 with the L-11 76.2mm gun

Although the Defence Committee under the Council of People's Commissars had resolved to equip the new T-34 with the 76.2mm F-32 gun as early as May 1939, it transpired that the first weapon to equip production T-34s in 1940 was the 76.2mm L-11 weapon, which had actually been decommissioned in that year. The Leningrad Kirov Plant, which had been charged to mass-produce the F-32 weapon, failed to do so, with those few being produced being employed by them in their own heavy KV-1 tank. In consequence, the only other weapon to hand with which to equip the first T-34s now coming off the production line at Plant 183 in Kharkov was the less-than-satisfactory L-11 weapon. As these were being held in artillery storage, they had to be prepared for installation in the T-34 before they could be used. The L-11 Model 1938/39 found its way into the first 453 T-34s that were produced and thus all came from the Kharkov production line. Almost all of the T-34s lost in the opening months of Barbarossa were equipped with this gun, with none appearing in German photographs beyond the autumn. The first T-34 evaluated by the Germans at Kummersdorf mounted this weapon.

With a length of 30.5 calibres, its performance was very similar to that of the German 75mm KwK37 L/24 mounted on the Pz.Kpfw IV. Whereas the muzzle velocity of the Russian tank was greater at 612mps (metres per second), that of the German 'heavy' tank was 450mps. The L-11 was problematic to operate – it was prone to jam and it was difficult to aim with accuracy. Aiming the weapon was achieved by using either a TOD-6 or a PT-6 panoramic periscope. The weapon had an elevation of 25 degrees and could be depressed to 5 degrees. It had a rate of fire of one to two rounds per minute.

Armament of the T-34/85

The need for a new gun in the T-34 is covered in another section of this work. What is addressed here, albeit in much less detail than that for the T-34/76, are the changes to the tank brought about by the addition of a new turret carrying the new, heavier and more powerful 85mm main gun. The information given is to cover the T-34/85 equipped with the Zis-S-53 gun – the main production variant.

The 85mm gun was much bigger than the 76mm F-34.

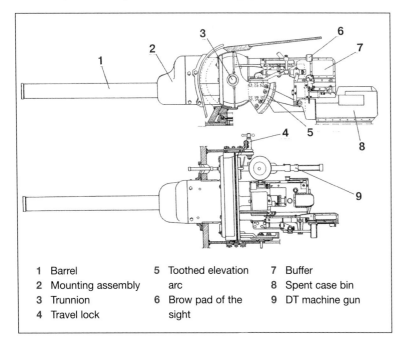

BELOW The L11 gun.
(Pen & Sword)

1 Barrel
2 Mounting assembly
3 Trunnion
4 Travel lock
5 Toothed elevation arc
6 Brow pad of the sight
7 Buffer
8 Spent case bin
9 DT machine gun

RIGHT These three interior shots are taken from the inside of the T-34/85 at Shrivenham. Dominating this image is the pad for the forehead of the gunner who looked through the monocular TShU-16 telescopic sight. To his upper left is the MK-4 periscope. Below the gunsight is the power traverse mechanism for the turret. To the gunner's right is the main gun breech block.
(All this page Dick Taylor)

CENTRE A more complete view of the inside of the turret allows a look in greater detail. As the T-34 was never fitted with a turret basket, the three seats for the commander, gunner and loader were attached to the turret ring. The pale green box in the top foreground is part of the radio equipment. The red leather seat resting against the turret side is the commander's seat, which would be lowered when the crew were inside. He stood on this to see out of the turret cupola when he wished to use his 'eyeball Mk.1' sight.

BOTTOM On the immediate right are empty clip holders for four 85mm shells, while the four circular containers are the spare ammunition magazines for the co-axial DT machine gun that can be seen to the right of the gun breech. The loader's seat hangs from the turret ring. The handle below the machine gun magazines is the turret traverse hand lock. One has to wonder how in combat in both the T-34/76 and the T-34/85 the gunner managed to reach down and pull free the shells stored in wooden containers on which the crew stood. The space in the later T-34 is not excessive and it must have been extremely difficult to move about. In the two-man turret in winter time, wearing thick sheepskins for warmth, it must have been nearly impossible. Small wonder that a Soviet officer said that Soviet tankers (unlike US soldiers) did not sleep inside their tanks – referring to the M4. The accommodation in the T-34 of any model was Spartan and functional. It also makes far more understandable the observations of Soviet tankers that the tarpaulins they were issued with and kept them dry at night when sleeping outside their tank were a carefully guarded treasure.

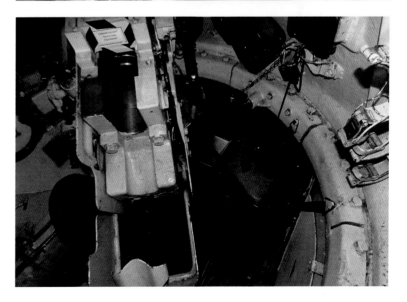

T-34 with the F-34 76.2mm gun

From the point of view of the Kharkov plant, the L-11 was an interim weapon which they had perforce to employ and wished to replace as soon as possible. The design bureau at Plant No. 9,2 overseen by V.G. Grabin, had been responsible for the design of the 76.2mm F-32 weapon and in 1940 he implemented a programme for improving it. The weapon that transpired was the gun that would equip the T-34 after it had replaced the L-11 on the production line and would be employed on all T-34s until it in turn was replaced by the larger 85mm calibre towards the end of 1943.

A mock-up of the new F-34 weapon was taken to the Kharkov works at the behest of the plant director, Maksarev, where it was deemed to be a most satisfactory fit for the T-34. It was then brought to the attention of the Defence Committee who ordered it to be installed in a BT-7 for further testing. Not surprisingly this resulted in the BT-7 incurring more than a few stress cracks – neither hull nor turret being engineered to cope with a gun that powerful. It was also tested in the A-34/2 tank with success. Although the commission did not sanction production of the weapon, it was recognised that it was superior to both the L-11 and the F-32, and could be employed in the turret of the T-34 without requiring any increase in the size of the turret ring.

The decision to place the F-34 in mass production was taken in December 1940, with manufacture beginning in early 1941. Even though the Gorki factory started delivering the F-34 to Kharkov in February 1941, the formal commissioning of the weapon for production

BELOW T34/76 cutaway side elevation, layout of T-34 mod 1942. *(Pen & Sword)*

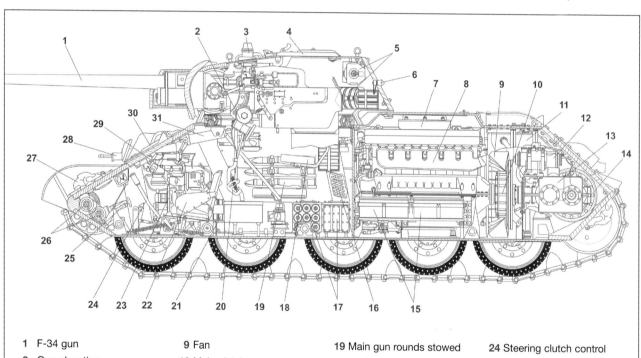

1. F-34 gun
2. Gun elevation mechanism
3. PT. 4-7 periscope
4. Roof hatch cover
5. Machine-gun magazines stowed in turret bustle
6. Firing port plug
7. Air filter unit
8. Engine
9. Fan
10. Main clutch
11. Rear fuel tank
12. Starter
13. Gearbox
14. Steering clutch
15. Batteries
16. Coil spring
17. Ammunition racks
18. Commander's seat
19. Main gun rounds stowed on the wall of the fighting compartment
20. Electrically actuated trigger pedal
21. Bow gun ammunition stowage
22. Driver's seat
23. Gear lever
24. Steering clutch control lever
25. Fuel pump
26. Compressed air bottles
27. Towing hook
28. Machine-gun mantlet
29. Radio set
30. Driver's hatch cover
31. Counterbalance of the driver's hatch cover

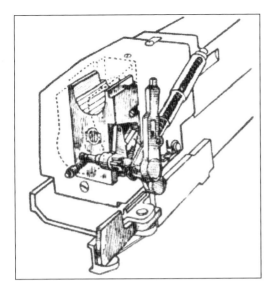

did not take place for a further five months! Nonetheless, T-34s with the F-34 installed began leaving the Kharkov factory in March 1941 and made their appearance on the battlefield alongside the earlier F-11-equipped machines in the summer/autumn battles in 1941.

The F-34 in detail

A very detailed analysis of the F-34 gun was carried out by the British School of Tank Technology as part of its much wider evaluation of a T-34/76 supplied by the Russian government for that purpose in late 1943. What is presented here is derived from that document and covers the main features of the weapon. If the reader is interested in the minutiae of the report (and it is as comprehensive as one is likely to find on this variant of the T-34), then details of the publication can be found in the bibliography.

It was noted that in the Russian-language handbook supplied, it referred to the F-34 as Models 1940, 1941 and 1942. The gun on the tank provided for evaluation was numbered 11141 and was presumed to be a 1942 model – the year of manufacture. The present tense used in the document has been retained.

Dimensions of the F-34 cannon

Length of rifling: 2,546mm
Length of chamber: 410mm
Length of bore: 2,956mm
Length of piece: 3,168mm
Weight of piece: 455kg

Details of the structure of the F-34

The gun is of a **Monobloc construction**, with a detachable breech ring secured by a locking ring. Two securing rings round the chase, retained by locking rings, carry the recoil cylinder and guides underneath the gun. The **breech mechanism** consists of a falling wedge breech block, with a hand or semi-automatic operation. Percussion cap firing is employed, the breech block carrying a readily removable striker and mechanism.

The gun cradle is of cast and welded construction, the lower portion of which is shaped in the form of a 'U'. The upper edges of the sides are formed into guide rails for the recoil of the piece. Two trunnion bearing castings are welded to the cradle with the trunnions themselves carried in two internal cheek plates welded to the turret front wall. The internal plate is bolted to the trunnion bearing castings. The left one carries the telescope and the operating arm for the periscopic sight object prism, and the right the co-axial machine gun cradle. The left-hand cheek plate carries the gearbox and hand wheel for the elevating gear. Elevation of 30 degrees, depression of 3 degrees and the total arc is of 33 degrees. It was noted that depression has been sacrificed in favour of a lower turret height and larger angle of elevation. The elevation angle of 30 degrees is abnormally large but acquired at the expense of depression.

The gun is centrally mounted in the turret (and was on all the different T-34/76 turrets) and

ABOVE LEFT AND ABOVE Breech mechanism.
(Nik Cornish)

is of a rear trunnion type and formed by two internally projecting cheeks welded to the inside of the front of the turret wall and turret ring. These carry two trunnions secured by brass caps. On the left of the cradle are carried the folding deflector guard, a small capacity [eight shell cases] empty cartridge bag, the externally toothed sector of the elevating gear, the firing gear for the 76.2mm gun, the brow pad for the telescope and the recoil indicator. The semi-automatic cam is on the right of the cradle. The rear deflector plate carries a fibre pad against which the empty shell cases are thrown on ejection.

The **recoil system** for the gun comprises a hydraulic buffer and hydro-pneumatic recuperator on the left, carried underneath the gun in eyes formed in the lower part of each securing ring. The piston rods are stationary and are nutted to the cradle cap. The recoil indicator on the left of the cradle is graduated from 310to 390mm with 'Stop' being at 390mm. The Russian handbook gives the normal recoil of the gun as being from 320 to 370mm and metal to metal recoil as 390mm.

As to the **balance** of the weapon – as it is in a rear trunnion mounting it is considerably muzzle-heavy. This is counteracted by means of four cast-iron blocks of 648cu in total volume/168.5lb approximate weight, bolted to the underside of the cradle. These are not fully effective and even with the blocks fitted the piece is still muzzle-heavy.

Hand and foot firing are provided for both the main gun and the co-axial machine gun. The foot firing for both guns is done from two pedals provided with return springs, and mounted on either side of a vertical pillar bolted to the left-hand trunnion mounting. The firing pedal for the 76.2mm gun is on the left and the machine-gun pedal on the right. The pedals are connected to the hand triggers on the guns by flexible cables. Foot rests are also provided on the pillar above the pedals. The hand firing gear on the 76.2mm gun consists of a spring-loaded lever which presses on a plunger passing through the left-hand breech ring side plate.

Hand **elevating gear** is fitted. The hand wheel is mounted on a longitudinal axis at about 30 degrees to the horizontal. It is on the gunner's right side and is mounted on the left trunnion bearing. It took 23¾ turns of the hand wheel to cover the full arc of 33 degrees. It was noted that there is considerable play in the hand wheel and its operation is jerky and not easy and for the gunner his position is cramped in consequence of which he would tend to catch his knees while elevating the main gun. In the case of the turret traversing gear, hand and power traverse through 360 degrees

BELOW F-34 with telescopic sight.

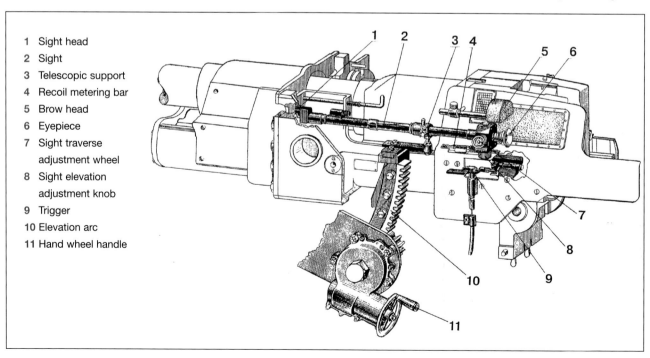

1 Sight head
2 Sight
3 Telescopic support
4 Recoil metering bar
5 Brow head
6 Eyepiece
7 Sight traverse adjustment wheel
8 Sight elevation adjustment knob
9 Trigger
10 Elevation arc
11 Hand wheel handle

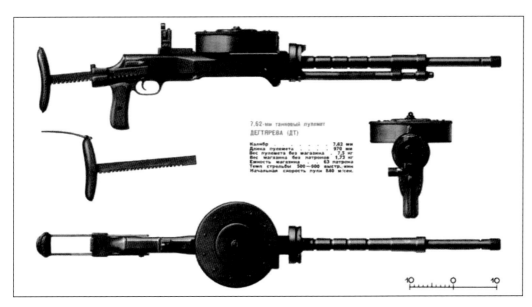

LEFT 7.62mm machine gun.

is provided. Both hand and power traverse employ a common gearbox, which consists essentially of a single epicyclic train. The sun pinion is driven by the motor, the planet carrier by the hand wheel, and the annulus drives the rack pinion.

The **power traverse** is merely a three-speed switching gear – not a laying gear – controlled for speed and direction by a small hand wheel on top of the traverse motor casing, having three positions in either direction. There is no positive indication of the three positions, but a catch automatically locks the wheel every time it is turned to 'neutral'. Power to the power traverse is provided by 20V 1,350W motor which the handbook stated could generate a maximum of 5,800rpm at 110amps and is mounted on the left of the gunner. When turned to the left the turret would traverse a full 360 degrees in 13.8 seconds and when moved to the right a full traverse would take fractionally less time – it being 13.6 seconds.

Selection of the alternative hand traverse mechanism sees the gunner employ a 4in radius hand wheel on his left which is placed on a horizontal axis. The hand wheel is not pivotally mounted on the wheel and it was noted by the evaluating team that considerable discomfort would be caused to the gunner when traversing the wheel quickly. It is awkwardly placed, being too far back for operation by the left hand, thus entailing the use of the right hand across the body (this was continued in the T-34/85). It would take the gunner 390 turns to traverse the turret through 360 degrees. It was also noted that there is a considerable play and backlash in the gearing, making accurate gun-laying difficult and quite significantly it makes indirect fire impossible. A traverse lock, of a screw type, is incorporated in the front offside turret ring clip, in front of the loader.

Co-axial machine gun

The same weapon was carried as a co-axial machine gun in the T-34/76 and in the T-34/85. It is a 7.62mm Degtyarev light machine gun. It is gas operated and magazine fed from drum magazines holding 63 rounds. It can be dismounted and employed as a ground light machine gun employing the bipod used in the stowage. It has a light barrel although there are no spares carried and in any case the barrel cannot be changed once it is mounted. The magazines appear to be extremely simple and efficient in operation. A deflector bag is fitted under the gun. The weapon mounting is of a ball-type fitting. It is held in a carrier bolted and welded to the right-hand side of the 76.2mm cradle. An aperture for use with the open sight on the gun is provided in the mantlet, and may be closed by a rotating shutter operated by a handle above the gun. The gun may be fired by the gunner from the right-hand foot pedal, or by the loader from the trigger on the gun. The foot-firing pedal is connected by a Bowden cable to a lever, with a spring return, mounted on the MG trigger guard, which operates the trigger of the gun.

The auxiliary machine gun is of the same type as the co-axial weapon and is mounted in an armoured hood welded to the offside of the glacis plate. The mounting carries an external mantlet and an integrally cast-iron jacket welded to an internal armoured ball which is similar to that in the co-axial machine gun. The inner ball is retained in the hood by means of a horseshoe-shaped ring, bolted to the front of the hood. The weapon can be traversed in the following fashion: elevation -20 degrees, traverse left and right -15 degrees, depression -6 degrees. The gun is fired by a trigger on the weapon itself and its movement controlled by the pistol grip. A travelling lock is provided at 8 o'clock in the ball carrier.

Sights

The main gun and the co-axial machine gun were sighted by means of either a Periscopic Dial Sight PT. 4-7, or a Cranked Telescopic Sight, Type TMFD.

Periscopic Dial Sight (PT. 4-7)

This is a periscopic sight, with a rotating head, movable top prism and illuminated graticules. It is mounted in the front of the turret roof on the nearside. The head of the periscope may be rotated independently of the turret by means of a knob on a vertical axis under the sight body.

It may also be locked in the forward position by means of a spring-loaded knob on the left of the body. A scale is connected to the head which may be read through a window in the rear of the body above the eyepiece – this is graduated from 0–60 (at 6 degree intervals) and gives the angle in a horizontal plane between the line of sight and the axis of the bore of the 76.2mm gun. For sighting, the head must be locked at '30'. The object prism in the head may also be depressed and elevated with the gun by means of an adjustable linkage connecting an arm on the left-hand trunnion bearing to a rotating arm on the right of the sight body. The arm on the sight is geared to a vertical push rod, against which the prism is spring-loaded. Deflection of the cross-wires is possible by the knob on the left of the eyepiece, and range is put on the knob under the eyepiece. Three ranges are provided:

- Scale 'П' – MG – 0–1,000m in 200m.
- Scale 'Б' – AP – 0–3,600m in 200m.
- Scale 'О' – HE – 0–2,100m in 50m.

(In each case the zero is displaced to allow for negative jump.)

A deflection scale graduated from left and right from 0–32 in mm is provided under the range scales. Both the deflection and range scales are engraved on glass carried in the vertical part of the body, whereas the cross-wires are in the eyepiece body. Separate illumination of range and deflection scales, cross-wires and the external scale for the rotating top prism is provided, the current being taken from a junction box on the sight body.

A rubber eye-guard and brow-pad, the latter adjustable so as to allow the use of either eye, are fitted. The eye-guard is not an efficient light excluder. The periscope is secured in the turret roof by means of a rotating locking ring, of similar pattern to that of the MG mounting, which engages with lugs on the periscope body and two horizontal adjusting screws on the mounting.

The object prism assembly is replaceable, and one spare is carried with each instrument. The object end of the periscope can be covered by a hinged hood on the turret roof.

Telescope TMFD

This is a straight tube moving eyepiece telescope, the object end being offset 23mm upwards from the axis of the body by the erecting prism assembly. It has an illuminated graticule and three range scales:

- Left-hand (Г) – HE (old type) – 0–3,800 (in 200m)
- Centre (Д Г) - HE (streamlined) – 0–5,000 (in 200m)
- Right-hand (О) – MG – 0–1,400 (in 200m)

(In each case the zero is displaced to allow for negative jump.)

Range is put on by a milled knob under the eyepiece body. Above the range scales there is a deflection scale graduated left and right from 0–32 in mils. Deflection adjustment is given by the knob on the left of the eyepiece body.

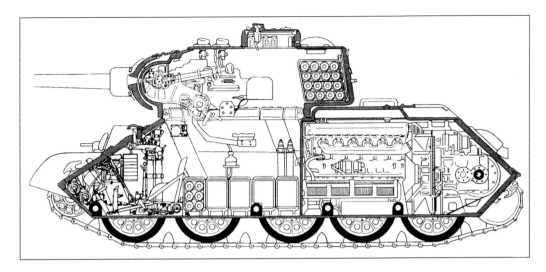

LEFT This cross-section of a T-34/85 shows how the 57 rounds of ammunition carried by this variant were stored differently to that of the 76 model. While the ammunition containers which formed the floor for the three men in the turret can be clearly seen and was the same as that on the earlier model, there is now storage for ammunition in the turret bustle. Normally 57 rounds of 85mm ammunition would be carried and would comprise of four types – the BR-365 or 'flathead' round, the BR-365 'sharphead' round, the BR-365R sub-calibre round and the O-365 HE round. *(Gennady Petrov)*

Separate illumination of range scales and cross-wires is provided:

- Magnification: ×2.5
- Field of view: 14° 30min
- Exit pupil diameter: 4.6mm
- Transmission of light: 39.2%

The telescope is mounted on the left of the 76.2mm gun with its nose resting on the left-hand trunnion casting. It is held secure by a compression spring shock absorber around the body which presses against a supporting bracket. The bracket is adjustable for zeroing – laterally by a traverse dovetail slide, and vertically by two adjusting nuts. A telescope is clamped in the bracket by a hinged strap and clamping nut. A rubber eye-guard is fitted, which the gunner used with his right eye.

Ammunition

The normal ammunition load carried by a T-34/76 was 77 rounds of 76.2mm rounds divided into 53 rounds of High Explosive, 19 rounds of Armour Piercing (tracer, with MA5 base fuse) and five rounds of Shrapnel. Six of these were carried on two racks of three which were mounted on the left wall of the fighting compartment and were retained by quick-release metal straps. A further three were mounted in the same fashion but on the right wall of the fighting compartment.

The bulk of the ammunition totalling the remaining 68 rounds, were stored in eight bins on the floor of the hull and actually formed the floor of the fighting compartment. When not in use the bins were covered by rubber matting which was then moved to gain access to the shells. Those in the bins were held in place by shaped wooden rests, hinged to one side of the bin and resting on stops at the other. Six of the eight bins held nine rounds in three layers of three with the remaining two holding seven rounds each. In practice and under combat conditions these were difficult to access. It was noted that the lids fit tightly and thus took time and effort to remove. Neither the lids nor the rubber matting could be dispensed with, making the job of the loader in selecting the appropriate ammunition difficult when in combat.

Three types of ammunition were stowed in the T-34 on its arrival at the School of Tank Technology and in the following numbers:

- Armour Piercing (AP) (tracer with MA5 base fuse) – 19 rounds
- High Explosive (HE) (with KTM I Percussion fuse) – 53 rounds
- Shrapnel (Ball with 22sec T6 fuse) – 5 rounds

	Lengths of rounds	Weight of rounds
AP	614mm	21lb
HE	634mm	20lb 6oz
Shrapnel (fuse cap off)	702mm	20lb 4oz
Shrapnel (fuse cap on)	715mm	20lb 10oz

In the turret used on the T-34 in 1941 and early 1942, a maximum of 77 main gun rounds could be carried but with the introduction of the new hexagonal turret in 1942 and models produced thereafter, up to 100 rounds could be carried.

HOW DID THE GERMANS VIEW THE T-34?

The most significant pointer to the profound impact the T-34 had made upon the Germans can be seen in that within a few months of the invasion it was recognised that the T-34 had rendered the Panzer III, the most important tank in the Panzerwaffe, obsolescent. Discussions concerning the post-Barbarossa plans for the Army held in early July had envisaged that the Panzer III would continue to be the main battle tank of the Army, with 7,992 being required for the 36 tank divisions deemed necessary 'for future operations' after Barbarossa. Shortly thereafter, this plan was deleted and by the end of 1941, the Panzer III was marked down for replacement with a new medium tank.

Although hundreds of T-34s were found abandoned by the advancing Germans, it took until December 1941 before an example arrived at the Kummersdorf *Kraftfahrerversuchsstelle* (motor vehicle test facility), thus permitting a proper evaluation of the type. In charge of the process was the head of the testing facility, Dr. Ing H. Knoppeck. Although it was later joined by others, the initial example of an early T-34/76 with the short 76.2mm gun was studied with great urgency. The conclusion was that in terms of its protection, mobility and firepower, the T-34 was superior to both the Panzers III and IV. Its wide tracks generated a lower ground pressure and thus made for a much better cross-country performance than the narrower tracks of the Panzers. In the final report detailing the evaluation penned by Colonel Dipl. Ing Esser, it was stated that the T-34 was a good design and that in the circumstances, the most sensible expedient would be simply to 'copy it'.

However, 'copying' the T-34 was more than could be expected as national pride was at stake and the German tank industry believed it could do better. Nonetheless, a large number of design features of the Russian tank were appropriated by Daimler-Benz (DB) in its VK 3002 DB submission for a replacement for the Panzer III. Many of these were very much a departure from normal German tank design practice. The adoption of sloping armour for its design also fed through to all new medium and heavy tank proposals conceived, designed and built by the Germans before the end of the conflict. In addition, the engine to be installed was a diesel (German tanks hitherto employed petrol engines) and it was to be mounted transversely and not longitudinally. It was also to utilise a rear drive rather than the standard German practice of powering the drive sprocket at the front of the machine via a driveshaft connected to the transmission housed on the floor of the tank and which sat between the driver and the radio operator. It would employ a leaf spring suspension with interleaved road wheels and be more heavily armed with a long-barrelled 75mm L/70 gun (the same gun that would be employed on the MAN design that was eventually chosen). Both the DB and the MAN submission had wider tracks than those used

RIGHT Although it took until December 1941 for the first captured T-34 to be transported to the German Army's evaluation centre at Kummersdorff, the judgement of the team that examined it was that 'it should be copied' as soon as soon as possible.

ABOVE The significance of this picture lies in that the T-34 is standing next to a Panzer III, which at the time of Barbarossa was the German Army's standard medium tank. A decision by the Army to substantially increase production of the Mark III in July 1941 was quickly rescinded once it was appreciated that the T-34 had rendered it technologically obsolete. The upshot was the emergence of the Panther as the new medium tank for the German Army.

on the Panzers Mark III and IV. Although Hitler expressed a preference for the DB design, the Army opted for the alternative proposal.

In part, those characteristics of the T-34 that were not incorporated into either of the two proposals reflected those that the German evaluation team had identified as weaknesses or flaws in its design. In many ways, these had been identified by the Russians themselves in 1940, and were to have been addressed in the T-34(M). The biggest German criticism was directed towards the two-man turret. It was regarded by them as small and extremely cramped. It was, however, the fact that the tank commander had also to double-up as the gunner that meant his situational awareness declined as soon as he took his eyes away from his sight. From the outset, the Germans had included a three-man turret crew in their Mark IIIs and IVs, and this would continue to be a normative feature of all future medium and heavy tank designs. This made for a more sensible workload and permitted the tank commander a continual view of the battlefield either through his vision slits or with his head poking out of the cupola. While the Soviets had recognised the value of a three-man turret and a commander's cupola (once more a feature of the T-34(M)), the need to ensure maximum output in production militated against its introduction until 1943, and the later models of the T-34/76.

In many ways it was this factor, combined with the much poorer optics for aiming the 76.2mm gun, that permitted German tanks with their five-man crews to reduce much of the inherent superiority of the T-34 design. The sights in German tanks were recognised by the Soviets as being superior to their own right through to the end of the war. Indeed, from the German perspective, the quality of the optics in the T-34 was so poor they described the crew in the tank as being effectively 'blind'. The problem with the optics in the T-34 derived from the poor quality of their manufacture. It is also likely that as it was an early production model of the T-34 that was evaluated by the Germans, it would have been equipped with the crude polished steel mirror type.

The last factor was the lack of inclusion of radios as standard equipment – a matter which profoundly impacted on the ability of the Soviet tank arm to function effectively in the first two years of the conflict. This profoundly reduced command and control of units, severely reducing their effectiveness, and was a consequence of the limited development of the electronics sector of Soviet industry. By 1943, not only had this native industry improved but many radios fitted in Soviet tanks had their origin in US, British and Canadian factories, being supplied and delivered via Lend-Lease. While recognising the fine balance between armour, mobility and firepower represented by the T-34, it was those other factors (plus others addressed elsewhere in this book) that served to degrade them and thus greatly reduce the performance of the T-34 in combat. It is these factors that go a long way in explaining the very high losses of the Russian tank to the Germans in the first two years of the Eastern conflict.

Appendix 1

T-34 variants including SPGs

The T-34 was produced in a multitude of variants – far too many to explore in detail. What is offered here is a short gazetteer identifying as many T-34 variants and SPG (SU series) as possible. To begin with, the breakdown covers those machines produced between 1940 and 1945, followed by post-war Soviet and non-Soviet variants.

1939–45	
T-34/76 Model 1940	First production model, armed with L-11 76.2mm gun in a welded or cast turret. Built at KhPZ No. 183.
T-34(M)	Tank that would have replaced the standard T-34 had the war not broken out. Extensively redesigned, with a three-man turret, torsion bar suspension, more powerful engine.
T-34/76 Model 1941	Appeared in 1941 and fitted with cast/welded turret. Main gun was now F-34 76.2mm. Built at KhPZ (until evacuated) and STZ.
T-34 Model 1942	Appeared in 1942. It was mainly produced with a cast turret. Side armour increased to 45mm. New wheel and track design.
	New hexagonal turret introduced – cast or pressed.
	New hexagonal turret with cupola – cast or pressed.
	New drop-forged turret produced by UTZM.
T-34/57	Limited production of T-34 produced with Zis-4 57mm anti-tank gun.
OT-34	Flame-throwing tank. Hull machine gun replaced with flame thrower.
T-43	Development of 76mm armed T-34 with bigger turret. Abandoned in favour of 85mm armed T-34 with new turret derived from the T-43.
PT-34 Mine roller	Mine roller mounted on T-34/76 1943.
T-34/85 Model 1943	Entered service in early 1944. New three-man turret armed with 85mm D-5T gun. It was equipped with a commander's cupola as standard.
T-34/85 Model 1944	Appeared in 1944. It was equipped with 85mm Zis-S-53 main gun. Commander's cupola moved further to rear on turret. Radio relocated from turret to front hull. New gunsight.
T-34/85 Model 1945	Appeared in late 1944. Equipped with larger turret cupola with single-piece roof hatch and electrically powered turret traverse. It was fitted with an electrically powered smoke system. A feature of this model produced at Plants No. 112 and No. 75 (formerly the KhPZ No. 183) were the separated mushroom vents on the turret roof.
SU-122 SPG	T-34 chassis with new fixed superstructure armed with 122mm M-30S howitzer and was built at URALMASH 1942/43.
SU-85	Tank destroyer built on T-34 chassis. It mounted an 85mm D-5T main gun and was built at URALMASH 1943/44.
SU-100	Tank destroyer mounting a 100mm D-10S main gun. Production began in late 1944 and continued post-war.
TT-34	Improvised recovery tank on T-34 chassis.

POST-WAR PRODUCTION AND REBUILDS IN THE USSR	
T-34/85 Model 1946	1946 production fitted with an improved V-2-34M diesel engine and new road wheels.
T-34 Model 1960	Rebuilt and modernised T-34. Now fitted with a better Model V-2-3411 diesel engine. Other improvements included smoke dischargers, infra-red headlamp, improved radio set and new air cleaner and generator.
T-34/85 Model 1969/ T-34/85 M	Modernised T-34. It was equipped with modern radio set, new road wheels and improved transmission system, night vision for the driver, external fuel tank support and an anti-ditching device under the rear hull.
OT-34/85	Flame-throwing variant of standard tank. ATO-42 flame thrower mounted in place of hull machine gun.
SPK-5	Battlefield engineering vehicle less T-34 turret that appeared in 1955.
SPK-5/SPK-10M	As above but with electrohydraulic-powered crane.
T-34-TO	Battlefield maintenance vehicle on T-34 chassis.

T-34S BUILT UNDER LICENCE AND OTHER NON-RUSSIAN DERIVATIVES OF THE T-34	
T-34/85CZ/T36	Licence-built T-34/85 produced in Czechoslovakia for domestic Czech Army and for export. Production began in 1951, ended 1955. Replaced by production of the T-55.
SU-100CZ	Licence production of the SU-100 in the Martin plant in Czechoslovakia.
MT-34	Czechoslovakian bridge-layer on T-34 chassis.
VT-34	Czechoslovakian recovery vehicle on T-34 chassis.
T-34/85 Polish manufacture	Licence production of 685 machines built between 1953 and 1954.
Type 58	Chinese T-34/85 production model.
Type 58 (59)	Chinese T-34/85 with Type 59 sights (the Type 59 being the Chinese copy of the T-54).
Type 58-I	Improved Chinese Type 58 fitted with revised commander's hatch and provision for heavy machine gun on the turret roof.
Type 58–II	As above, but fitted with additional commander's cupola and provision for two heavy machine guns on the turret roof.
Type 63	Chinese SP anti-aircraft gun mounting twin 37mm guns.
T-100/T-34-100	Egyptian tank destroyer, equipped with 100mm BS-3 anti-tank gun.
T-122	Egyptian SP gun fitted with 122mm D-30 howitzer.

Appendix 2

The T-44

In the summer of 1944 the Morozov design team at the Uralvagonzavod turned their attention to the design of a new tank to follow on from the T-34/85. Production of the latter was at full tilt, but it was recognised that its design potential had reached its limit. Indeed, Stalin had indicated as much in the autumn of 1943, stating that what was required was a new medium tank that would be superior to the German Panther. Herein lay the genesis of the medium T-44.

Allocated the designation Obiekt 136, the Morozov design drew on aspects of the T-34 for the new design of the new medium tank. By way of example, it retained the road wheels, the turret and armament as well as the engine of the older design but it also included a significant number of innovations, the most important of which was the adoption of a torsion bar suspension system, a feature incorporated on the abortive T-34(M) design of 1941. This was as much a recognition that the Christie suspension system was now passé as it was to do with permitting a radical reduction in the height of the new design. Additionally, mounting the engine in a transverse fashion allowed the turret to be placed in the centre of the hull. The hull sponsons of the T-34 were abandoned with the new hull design permitting the crew section to be more spacious internally. Qualification is needed here, as the extra space for the crew was only marginally greater than for the T-34/85 and much less than found on the US M4 medium tank, for example.

The armour protection for both the hull and turret was superior to that of the T-34, albeit on a smaller hull. The glacis, for instance, was a much thicker 90mm of armour sloped at 64 degrees – a significant increase in frontal protection when compared with the T-34's 45mm at a 60-degree slope. This was superior even to that of the German Panther's glacis which had 80mm of armour on a 55-degree slope. The hull machine gun was deleted and while the first two prototypes had a driver's hatch as part of the upper glacis, this was removed in the third, and placed on the front of the hull top. To accommodate this change, the turret was moved back a little on the hull top. Armour protection on the turret was also increased over that of the T-34/85, with the turret front and the gun mantlet being of 120mm thickness. The latter housed a main armament that was the same as that of the T-34/85 – the 85mm D-5T gun.

Although the T-44 was to evidence numerous problems, it was not actually these that led to its early production demise. What did for the tank was its gun. The Red Army was looking for a bigger calibre than the 85mm already being carried by the T-34/85. Indeed, the trialling of a number of T-34/85s, where the main gun had been replaced by the same 100mm weapon carried by the SU-100 tank destroyer, was a pointer to what they really wanted – a medium tank mounting a 100mm gun. As it was, there were problems with the T-34 being able to mount such a large weapon and it never proceeded to production. The fitting of the 85mm calibre had not permitted the T-34 to equalise fully with the Panther,

BELOW By 1944 it was acknowledged that the T-34 – even its 85mm-armed variant – was in need of replacement. With the war having turned decisively in favour of the Soviet Union, A.A. Morozov and his team at Nizhne Tagil turned to designing a new medium tank. The outcome was the T-44. This is the prototype with the driver's hatch built in to the glacis. *(All author)*

and with the introduction of the much heavier and more powerfully armed Tiger II with its KwK43 L/71 main gun in the summer of 1944, 85mm was no longer a calibre up to the task. It would be this that would doom the T-44, but much in its design would find its way into its replacement – the T-54. In the same fashion that a number of T-34/85s had been trialled with the 100mm D-10, a T-44 was also trialled with the same weapon. In neither case were the 100mm variants proceeded with. There was also an attempt to fit a 122mm gun as mounted on the IS-2 heavy tank into the turret of the third prototype. This also proceeded no further than the trials stage.

Production of the T-44 began at the former Zavod No. 75 engine factory in Kharkov, which had been converted into a tank manufacturing plant when the Morozov design bureau returned there in 1944. The first production machines were delivered to the Red Army in the late summer of 1944 with the first units being formed in September. Three Guards tank brigades, namely the 6th, 33rd and 66th, were selected to receive the new tank but they were thereafter employed to train up other units converting to the T-44. Although there have been rumours that the T-44 saw some service at the very end of the war in Europe, this was not the case. The design underwent some modifications and was in production through to 1947, with a total of 1,823 units being built. Unlike the T-34, the T-44 saw no service other than with the Red Army. That it did not replace the T-34/85, notwithstanding its superior design, had to do with its armament. Mounting the same 85mm weapon as the T-34, it was not perceived as representing a sufficient advance, especially as the Soviet High Command had made it clear that they wished to see the 100mm D-10 equipping their new medium tank. With a new turret and the necessary redesign of the hull to accommodate it, this would emerge in 1947 as the T-54.

RIGHT The T-44 was in production from 1944 through to 1947, but at 1,823 the numbers built were relatively small. It was a remarkable design but its Achilles heel was the size of its main gun. By 1945 the Red Army wanted a medium tank mounting a 100mm weapon, but this would come with the successor of the T-44, the T-54.

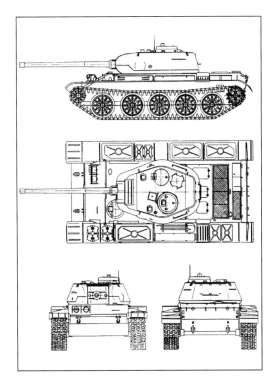

LEFT This three-view drawing shows the production model. The glacis is now clear, the driver's hatch has been moved to the top of the hull and the turret moved back to its centre. It has a torsion bar suspension and the engine is mounted transversely.

ABOVE This T-44 was fitted and tested with a 122mm gun. It was not accepted for production.

Appendix 3

The T-34 turrets (Mark Rolfe)

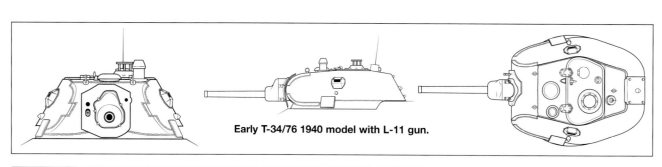

Early T-34/76 1940 model with L-11 gun.

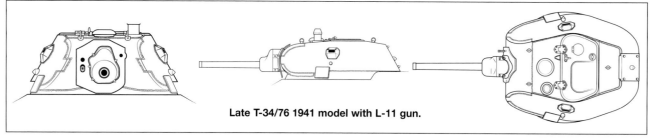

Late T-34/76 1941 model with L-11 gun.

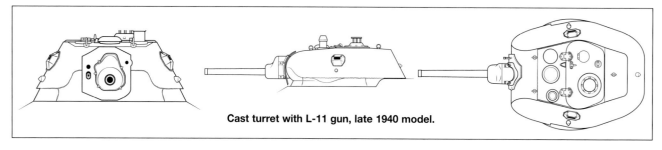

Cast turret with L-11 gun, late 1940 model.

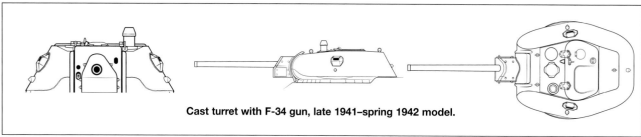

Cast turret with F-34 gun, late 1941–spring 1942 model.

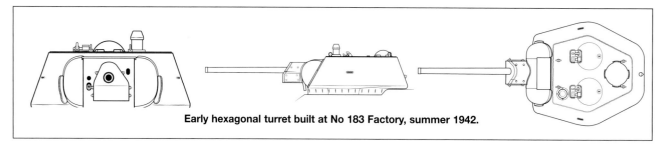

Early hexagonal turret built at No 183 Factory, summer 1942.

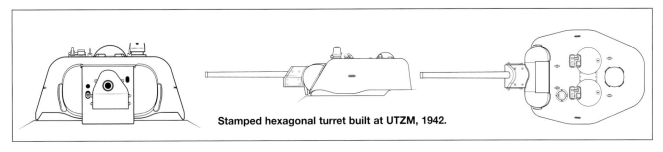
Stamped hexagonal turret built at UTZM, 1942.

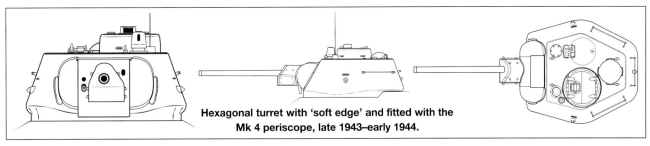
Hexagonal turret with 'soft edge' and fitted with the Mk 4 periscope, late 1943–early 1944.

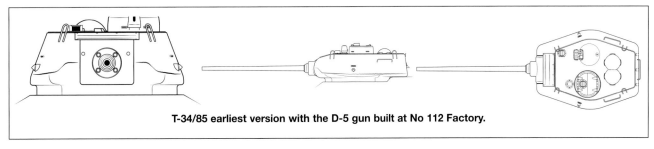
T-34/85 earliest version with the D-5 gun built at No 112 Factory.

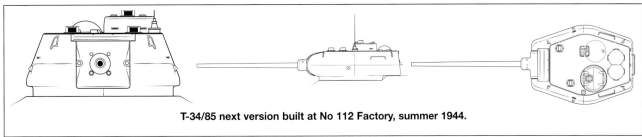
T-34/85 next version built at No 112 Factory, summer 1944.

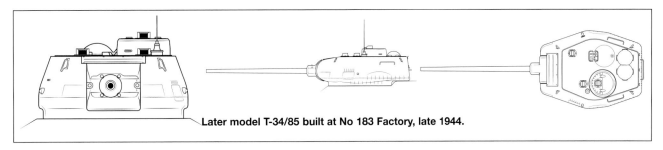
Later model T-34/85 built at No 183 Factory, late 1944.

Selected bibliography

Baryatinsky, Mikhail, *Light Tanks* (Ian Allan, 2006)

Baryatinsky, Mikhail, *T-34 Medium Tank* (Ian Allan, 2010)

Bessonov, Evgeni, *Tank Rider* (Greenhill Books, 2003)

Carell, Paul, *Hitler's War on Russia* (Harrap, 1964)

Drabkin, Artem and Sheremet, Oleg, *T-34 in Action* (Pen and Sword, 2006)

Erikson, John, *The Road to Berlin* (Weidenfeld & Nicolson, 1983)

Finkel, Meir, *On Flexibility* (Stanford Security Studies, 2011)

Glantz, David, *Stumbling Colossus* (University Press of Kansas, 1998)

Habeck, Mary, *Storm of Steel* (Cornell University Press, 2003)

Healy, Mark, *Zitadelle – the German Offensive against the Kursk Salient* (The History Press, 2008)

Jentz, Thomas L., *PanzerTruppen Volume 1* (Schiffer Military History, 1996)

Krivosheim, K., *Soviet Combat Casualties and Combat Losses in the Twentieth Century* (Pen and Sword, 1997)

Mawdsley, Evan, *Thunder in the East* (Hodder Arnold, 2005)

Michulec, Robert and Zientarzewski, Miroslaw, *T-34 Mythical Weapon* (Armageddon and Air Connection, 2006)

Milsom, John, *Russian Tanks 1900–1970* (Arms and Armour Press, 1970)

Moynahan, Brian, *The Claws of the Bear* (Hutchinson, 1989)

Orgorkiewicz, Richard, *Tanks – 100 Years of Evolution* (Osprey, 2015)

Raus, Erhard, *Panzer Operations* (Da Capo, 2003)

Stone, David (ed.), *The Soviet Union at War* (Pen and Sword, 2010)

Vollert, Jochen, *Panzerkampfwagen T-34–747 (r)*, (Tankograd Publications, 2013)

Vollert, Jochen, *Various Tankograd Gazettes* (Tankograd Publications)

Werth, Alexander, *Russia at War* (Barrie and Rockcliffe, 1964)

Zaloga, Steven, *Armored Champion* (Stackpole Books, 2015)

Zaloga, Steven and Grandsen, James, *Soviet Tanks and Combat Vehicles of World War Two* (Arms and Armour Press, 1984)

Zaloga, Steven, Kinnear, Jim and Sarson, Peter, *T-34/85 Medium Tank 1944–1994* (Osprey Books, 1996)

Zaloga, Steven and Ness, Leland S., *Red Army Handbook 1939–1945* (Sutton Publishing, 1998)

Zaloga, Steven and Sarson, Peter, *T-34/76 Medium Tank 1941–1945* (Osprey Books, 1994)

Index

AAD (Automobile and Armour Directorate) 25-27
Aberdeen Proving Ground, USA 100, 120, 122, 132-133, 135, 143
ABTU 25
Ammunition 26, 43, 103, 153
Angolan Civil War 111
Anti-aircraft guns 77
Anti-tank guns 45-48, 71, 76, 80-81, 91, 111, 123
Arab-Israeli Conflicts 1956 and 1967 110-113
  Six-Day War 1967 110-111
Aria, tankman Semyon 100
Armament (Weaponry) 26, 29, 31-33, 40, 46, 69, 76-81, 85-87, 89-91, 93, 102-103, 121, 128, 154, 148-153
  co-axial machine gun 151
  F-34 76.2mm gun 148-151
  L-II 76.2mm gun 146
Armour protection 26, 33, 46, 53, 75, 116, 120, 122-123, 128
  add-on-hull armour (AOHA) 123
  bedspring stand-off 94, 96
  shell-proof 24-30
  sloped 25-26, 33, 58, 84, 117, 127
  thickness 29, 85, 117, 122-123, 127
Artillery tractors 10-11
  Komintern 10
  Voroshilovets 11

Battle Fronts
  Belorussian 91, 94-95
  Central 73
  Eastern – throughout
  Leningrad 21
  Ukranian 86-87, 93
  Voronezh 73
Battle of Flers-Courcelette 6
Battles of Kharkov 58, 60, 67-68, 83
Battle of Kursk-Orel 49,68-69, 71-76, 79
Battle of Mtsensk 48-51
Bay of Pigs invasion, Cuba 111
Beria, 30-31
Berlin, Battle for and Fall of 90, 93-95
Bodnar, A.V. 101
Bolshevik Party 7
Bondarenko, 25-26
Break of Siege of Leningrad Museum 19
British Army 6, 17
  Experimental Armoured Force 11
  School of Tank Technology 6, 116, 149
  Tank Corps 11
British tanks and AFVs 11
  Carden-Lloyd tankettes 17
  Centaur 122
  Centurion 108, 110
  Challenger 122
  Charioteer tank destroyer 86
  Covenanter 122
  Cromwell 122
  Crusader 122

Mark IV 106
Mark V 7
Matilda II 83
Medium Mark IIA 17
Vickers
  6-ton 16-18
  Independent 20-21
  Mark III 20
Bryukhov, V.P. 88, 101
Burtsev, Alexander Sergeyevich 63

Camouflage and paint schemes 50, 65, 83, 110, 113
Captured tanks 7, 11, 82-83, 85-86, 111, 122, 125, 134
  Beutepanzer 82-83
Central Artillery Design Bureau 77
Chinese tanks, 106, 111
  Type 58 106, 111
  Type 59 106, 111
Christie, J. Walter 16-17
Citroën 17
Cold War 6, 106
Crew 26, 100-103, 120
  driver/mechanics 100, 120-121
  effects box 128
  headgear 100
  injured 61
  loaders 100-103
  radio operator/machine gunners 100-101, 120-121, 142, 147
  seats 142, 155
  tank commanders 100-102, 147
  training 88-89, 100
Cypriot National Guard 111

DPRK 125
Dzerzhinsky, Felix 53

Egyptian Army 110-111
Electrical system 128, 136-137
  accumulators 137
  rotary base junction 137
Engine 19, 26, 61, 89, 91, 103, 121, 131-137
  air cleaner 132-134
  cooling system 134
  electric starter 129
  exhaust system 134
  life 133
  oil and lubrication systems 62, 134-135
  V-2 diesel 28-30, 32, 34, 103, 131-137
Engine compartment 120, 128, 131
Equipment fitted
  flamethrower 77
  PT-3 mine roller 74, 93
  smoke generators 94-95, 128
Evacuation and relocation of tank factories 51-56, 62-63, 101, 123
Evolution of medium tanks 30

Fedorenko, N. 133
Finnish Army Museum 86
First World War 7, 11, 35

Five-Year Plans 11, 14
French Army 26
French surrender to Germany 34-35, 38, 45, 83
French tanks 11
  FMC-36 26
  FT-17 7
  FT-18 8
Fuel system and tanks 31, 45, 62, 66, 128, 130-131, 135-136
  injector equipment 134
Fuller, 23

German Army Foreign Vehicle Classifications 82-83
German invasion of Soviet Union 34-35, 38-57
German/Russian Tank School, Kazan 10, 16
German surrender 95
  division of Germany 96
German tanks and AFVs
  Ferdinand 72, 74
  Panther 51, 69-73, 76-77, 79, 85, 88-90, 92, 116; Ausf A 70; Aust D 116; Aust G 116
  Panzer 1 22
  Panzer II 39
  Panzer III (Pz.Kpfw.III) 34, 38-39, 46, 50-51, 56, 59, 71-72, 74-75, 123, 154, 155
  Panzer IIIG 33; IIIJ 60
  Panzer IV (Pz.Kpfw.IV) 39-40, 46, 50, 57-59, 68, 71, 74-75, 88, 155; IV F2 60; IVG 58; IVH 92; IVJ 92
  Panzer 35(t) 40
  Pz.Kpfw 38(t) 39
  Sturmgeschütz (StuG) III assault gun 40, 71, 92
  Tiger I 67-69, 71-74, 77, 79, 85-86, 91
  Tiger II 91
  VK 3002 DB 154
Ginsberg, Semyon 17
GKO (State Defence Committee) 26-27, 29, 52, 63, 77-78, 84, 86, 123
Glantz, David 35, 42
Gorlitsky, L.I. 80-81
Grabin, V.G. 77, 87, 148
Great Patriotic War (Second World War) 97, 103
Guderian, Gen Heinz 49, 51, 74, 89
GUVP (Main Department of the War Industry) 8

Hitler, Adolf 35, 38-39, 60, 72, 76, 155
  accession to power 16
Hull 57, 60, 86, 116-117, 120, 123, 148
  glacis plate 33, 85, 117, 120, 122-123, 127, 141
  interior 120, 142-143
  riveted 40
  welded 84, 120
Hungarian Uprising 107

Ilyich Metalworks 34
Iraqi Army 112
Israeli Defence Forces 110-111

Japanese Army 21, 24
Japanese Kwantung Army 95
Japanese surrender 96

Kalinovski, Col. K.V. 11
Kerichev, V. 86
Khalepski, I. 15, 17, 23
Kim Il-Sung 107
Kinchenko, radio operator Petr 100
Knoppeck, Dr Ing H. 154
Koniev, 94-95
Korean War 96-97, 106-110, 116, 122, 125, 134
  UN landings 109
Koshkin, Mikhail I. 25, 27-32
Kravchenko, 96
Krivosheev, historian G.E. 76, 92
Kubinka Proving Ground 26, 29-31, 33, 67, 77, 86
Kucherenko, N. 28
Kulik, G.I. 23, 31
Kutukov, Col Mikhail 49, 87

Langermann, Freiherr von 50-51
Lelyushenko, Gen D.D. 49
Lenin 7, 70
Liddell Hart, Basil H. 23
Limitations and problems of the T-34 33-34, 61, 84, 100, 132-133, 137
Luftwaffe 38, 48
  Stukagruppen 48

Maksarev, Yu. E. 53, 148
Malinovskii, R. Ia 24
Malyshev, V. 31, 56-57
Maryevski, driver A.V. 100
Military districts in USSR western frontier 41
Molotov, Soviet premier 27, 31
Morgunov, Maj Gen 45
Morozov, A.A. 10, 27-29, 32, 54, 84, 131, 158
Moscow Central Museum of Armed Forces 20
Mostovenko, historian 26
MPLA, Angola 111

Nabutovsky, M.A. 67
NKO (The People's Commissariat for Defence) 32, 35, 146, 148
NKVD 30, 53, 133
Non-Aggression Pact with Nazi Germany 30, 33, 35
North Korean People's Army (NKPA) 81, 107-109, 125
North Vietnamese Army 111

Operation Bagration 90
Operation Barbarossa 35, 38, 45-46, 48, 51
Operation Citadel 71-76
Operation Kremlin 58

Operation Typhoon 48
Ordzhonikidze, Sergo 25

Panzerwaffe 39-40, 46, 69, 75
Pavlov, Gen D.G. 22-23, 25-26, 31
Performance 26, 32
  German view of 154-155
  manoeuvrability 46, 58
  mileage between overhauls 133
  range 26, 32, 89, 133
  speed 26, 58
Periscopes and telescopic sights 88, 147, 150, 152
  periscopic dial sight (PT. 4-7) 152
  telescopic TMFD 152
Petrov, F.F. 77, 87
PLA (People's Liberation Army), China 81, 106
Politburo 29
Preserved tanks 8, 18-21, 86, 116-117, 125-129, 147
Production 19, 31-34, 52-53, 56-57, 61-63, 66, 70, 72, 76, 78-79, 81, 86-87, 90, 97, 100, 106, 120, 131
  exports 6, 81, 106, 110
  in wartime 62-63
  post war and rebuilds 157
  relocation 51-53
  under licence 106, 110, 116, 125, 146, 156
  unit cost 70
  1940-45 54
Propaganda 7, 21, 49

Race from Kharkov to Moscow 30
Radios (wireless) and intercoms 100-101, 120, 141-143, 155
Raus, Erhard 68
Red Air Force 38
Red Army (RKKA) – throughout
  Mechanised Corps 11, 42-44, 46, 57
  Polish forces 80
  Tank Corps 29, 57
Reichswehr 7, 10, 16
Rhee, Syngman 107
Riabyshev, Maj Gen D.I. 43
Road (bogie) wheels 49, 64, 82-83, 86, 129-131
  locomotive-style 56, 60
  spider 95
Rokossovksy, 94
Rotmistrov, Gen Pavel 77
Royal Military College of Science, Shrivenham 116, 122, 125, 128, 147
Running gear – see also Road wheels 120
  idlers 140
Russian revolution 7
  White forces 7
Russo-Finnish War (Winter War) 1939/40 21, 35
Russo-German War 18

Sagun, Ivan 74
Schneider 17
Self-propelled guns on T-34 chassis 78-81, 92
  SU-85 79-80, 90; SU-85M 79
  SU-100 80-81, 111, 113, 129
  SU-122 (SU-35) 78-79, 95
Shaposhnikov, Head of Red Army Staff 14, 23
Shestopalov, Maj Gen N.M. 44
Škoda 17

Smirnov, tank driver Mikhail 99
Somua 17
Sokoolnikov, G. 7
South African Defence Forces 11
South Korean Army 107
South Vietnamese Army 111
Soviet tank designs
  A-20 (BT-20) 26-27, 29-30, 32;
  A-20G 28, 30
  A-32 28-30
  A-34 prototypes 29-31; A34/2 148
  BT series 17, 20, 22, 24-26, 28, 32, 34, 39, 42, 45-46
  BT-1 18; BT-2 18-19; BT-5 19, 22; BT-7 19, 32, 44, 103: BT-7M 19, 28; BT-7-IS 26; BT-IS 26, 28-29
  IS-2 70, 94
  KS 7-8
  KV 6, 29, 34-35, 38, 43; KV-1 32, 38, 40, 44-46, 50-51, 57, 63-64, 71, 146; KV-2 32, 38, 44-46
  MS-2 8
  MS-3 8
  OT-34 77
  SMK 29
  SU-76 light tank destroyer 109; SU-76M 109
  T-18 (MS-1) 8
  T-12 10-11, 15
  T-24 10-11, 15
  T-26 16, 18, 20, 24, 32, 42, 45; T-26A 18; T-26 B-2 22
  T-28 20, 45; T28A 21; T-28B 21; T-28C 21
  T-32 21, 29
  T-34 Models and variants 88, 156-157
    Model 1940 30, 40, 44-45, 123, 138
    Model 1941 32, 46, 48-50, 82, 85, 123
    Model 1942 73, 116, 123, 148
    Model 1943 64, 67, 82-83, 87, 102, 133, 136
    Model 1945 125
  T-34/76 6, 32, 34, 58-59, 67-68, 70, 77-79, 82, 86, 90, 92-93, 95, 97, 100-101, 116, 123, 130-131, 136-137, 153
  T-34/85 6, 70, 80-81, 83, 85-97, 102, 106, 108, 110-113, 116, 125, 128-130, 134, 136, 139, 147, 151, 153, 158
  T-34(M) programme 33-34, 77, 83, 124, 155, 158
  T-35 20, 45; T-35A 21
  T-37 32
  T-38 32
  T-40 32
  T-43 34, 54, 84-85
  T-44 54, 106, 124, 158-159
  T-50 32
  T-54 106, 110
  T-55 106, 110
  T-60 57
  T-70 71
  T-90 131
  T-100 29
  T-111 (T-46-5) project 27-28
  V-26 (Vickers 6-ton) 17
Spanish Civil War 21-22, 24-25, 45
Specifications T34/76 and T34/85 120
  engine 131
  weight 85
Speer, Albert 69, 76

Sprockets 124
Stalin, Iosif – throughout
  decision to stay in Moscow 51
  purges 23, 25-27
STAVKA (Soviet High Command) 42, 65
  'No Step Back' pronouncement 65
Steering 140-141
Steup, Lt 46
Supplies 43
  PO (petrol, oil and lubricants) 43
Suspension 11, 26, 92, 116, 122, 124
  Christie 17, 26, 28, 33-34, 84, 124
  cranks 130-131
  springs 122, 124, 142
  torsion bar 33-34, 84, 124
Sytnik, Col V.V. 73
Syrian Army 111
Syrian Civil War 6, 106, 113

Tank factories (German) 17, 69
  Daimler-Benz 17, 154
  Krauss-Mafei 17
  Krupp 8, 10, 17
  Krupp-Gruson 69
  Linke Hoffman 17
  MAN 69
  MNH 69
  Rheinmetall 10, 17
Tank factories and design bureau (Soviet) 51-55, 124
  Armament Plant 9 77
  Kharkov Zavod (KhPZ) 10, 15, 19, 21, 25-26, 28-32, 51-53, 131, 148
  Kirov Plant, Leningrad 20
  Krasnoe Sormovo (KS) 7
  STV 123
  STZ, Stalingrad 31-32, 34, 48, 51, 54, 56-57, 60, 65-66, 117, 123
  UZTM (URALMASH) 55, 64, 66-67, 77
  Zavod (Plant) No. 100 ChKZ (Tankograd) 54, 63, 66
  Zavod No. 112 (Krasnoi Sormovo), Gorki 51, 54, 64, 66, 83, 86-87, 89, 90, 117, 123, 125-126, 128-129
  Zavod No. 174 Omsk 55, 64, 66, 90, 125
  Zavod No. 177 Vyska Works 123
  Zavod No. 178 Kulebak Works 123
  Zavod No. 183 25-26, 30-34, 48, 51, 53, 67, 82-84, 95, 97, 106, 123
  Zavod No. 183 (UVZ/UTZ) 53-54, 56, 90, 125-126
  Zavod No. 264, SSZ 123
Tank losses 41-46, 48, 53, 57-58, 64-66, 70, 72, 75-76, 92-97, 108-109
  causes 1941-2 59, 155
Tank Museum, Bovington 116, 125, 128-129
Tank roles 23
  infantry support 23, 29, 47
  suppression of uprisings 106-107
Tarshinov, M. 10, 26
Testing and trials 31-32
Timoshenko, Defence Minister Marshall S.K. 35, 42
Towing and lifting eyes 89, 141
Tracks 56, 129, 140-141
  adjustment 141
  pins 129

Transmissions 10, 26, 84, 100, 117, 121, 129, 133, 137-141
  clutch 137
  differential cover 129
  final drive 139-140
  gearbox 100, 129, 138-140
  Maybach 10
  spare carried 138
Treaty of Rapallo 8
Triandafillov, V. 11
Trotsky 7
Tsyganov, N. 26, 28
Tukhachevsky, Gen Mikhail 11, 14, 23, 90
  execution 14, 23
Turrets 26, 31, 34, 60, 85-86, 91, 97, 101, 103, 116-117, 121, 125-127, 147-148, 160
  castings 126-128
  commander's cupola 83-85, 88, 93, 102, 126
  Gaika (hex-nut shape) 64, 67, 77, 82-83, 97, 102, 130
  hatches 40, 94, 121, 127-128
  multi 20-21
  welded 58

UMM (Directorate of Motorisation and Mechanisation) 15
US Army 16, 109, 116, 122
US Cavalry 16
US Marine Corps 109, 116
US tanks and AFVs
  Christie M1928 17-18; M1931 T3 16, 18; M1940 17
  M4 series 108, 134, 147; M4A2 Sherman 100; M4A3E8 108-109
  M-24 109
  M-26 Pershing 108-109
  M-46 108, 110
  M-47 110
  M-48 110

Vaishev, P. 28
Vasil'evich, Maj Gen B.S. 43
Verkhnyaya Pyshma military equipment museum 8, 19
Versailles Treaty 7-8
Vickers 17
Victory parades 95, 97
Vietnam War 111
Vigor, P.H. 70
von Dirksen, German ambassador Herbert 15
von Hammerstein-Equord, Gen Kurt 16
Voroshilov, Defence Minister K.E. 14-15, 17, 22, 27, 31, 35, 55
Vosnesensky, Nicholas 52

Warsaw Pact 106
Weaponry – see Armament
Wehrmacht 22, 38, 69
Werth, Alexander 62
Wheel-cum-track machines 18, 25-26, 29
Wireless – see Radios
Workers in Soviet industry 63

Yemeni Civil War 6, 81, 106, 113
Yugoslav Army 112
Yugoslav Civil War 81, 112

Zaltsman, Col I.M. 56
Zhukov, 23, 51, 77, 94-95